Stem Cells Made Easy

Jon Adams

Copyright © 2024 Jonathan Adams

All rights reserved.

ISBN: 9798326314710

CONTENTS

1 The Stem Cell Universe .... Pg 6

2 The Lifecycle of a Stem Cell .... Pg 17

3 The Blueprint of Life .... Pg 28

4 Regeneration and Healing .... Pg 40

5 Stem Cells in Research .... Pg 51

6 Ethical Considerations .... Pg 62

7 Stem Cells and Modern Medicine .... Pg 76

8 Future Horizons .... Pg 89

9 The Laymans Laboratory .... Pg 102

INTRODUCTION

Welcome to 'Stem Cells Made Easy,' a comprehensive guide designed to navigate you through the fascinating world of stem cells from A to Z. This book is tailored to illuminate one of science's most profound topics using a tapestry of deep analogies and vivid examples, making complex concepts not only palpable but also engaging.

From the basics of what stem cells are to the intricate details of how they can be used in medicine and research, 'Stem Cells Made Easy' peels back the layers of scientific jargon to reveal the core principles at play in regenerative biology. We'll embark on a journey exploring how these cellular marvels hold the key to unlocking new medical treatments and understanding human development.

You can expect to gain a solid foundation in stem cell knowledge, enriched with relatable analogies that lend clarity to advanced subjects. Each chapter is a building block, leading you confidently through the intricacies of stem cell types, their natural behaviors, and the cutting-edge techniques for manipulating them in the lab.

This book is a bridge between lay interest and professional-level comprehension, providing insights into the latest scientific findings while remaining rooted in everyday language and imagery. Whether you're a student, an educator, a healthcare professional, or a curious mind driven by the wonders of biology, 'Stem Cells Made Easy' promises a captivating and educational experience that will not only educate but inspire.

CHAPTER 1 THE STEM CELL UNIVERSE

Welcome to "The Stem Cell Universe," the first chapter where we explore stem cells, the foundational building blocks capable of developing into various cell types in the body. These cells are essential as they provide the raw material for both the growth and repair of tissues and organs. Their ability to differentiate into specialized cells makes them crucial in medical research, offering potential treatments for a range of diseases. Understanding stem cells is fundamental to advancing medicine and offers hope for regenerative therapies that could transform healthcare. This chapter will ground you in the basics of stem cells and their pivotal role in biological development and medical innovation.

Stem cells are unique cells with the remarkable ability to develop into many different cell types, from muscle cells to brain cells. They serve as the body's repair system by dividing without limit to replenish other cells, providing new cells for the body as it grows or heals from injury. Two key qualities define stem cells: self-renewal and potency. Self-renewal means they can replicate themselves many times over, while potency refers to their capacity to differentiate or transform into a variety of cell types.

There are multiple types of stem cells, each with varying levels of potential. For example, embryonic stem cells have the highest degree of potency, able to generate every cell type in the body. Adult stem cells, while more limited, are still vital for tissue repair and maintenance. Induced pluripotent stem cells, on the other hand, are adult cells that have been genetically reprogrammed to behave like embryonic stem cells. This ability to return to a more versatile state has significant implications for personalized medicine, as it could allow for the creation of patient-specific cell lines for therapy or research.

To visualize the concept, consider a library full of unclassified books. Embryonic stem cells are akin to a master index that can direct you to any book in the library, while adult stem cells are more like specialized guides that can only lead you to certain sections. Induced pluripotent stem cells are like taking a history book and magically giving it the ability to guide you to any genre. Each serves an important purpose in the intricate ecosystem of cellular function and medical applications, with the potential to treat a vast array of human diseases and conditions.

Stem cells possess two unique capabilities: self-renewal, the ability to divide and reproduce themselves indefinitely, and differentiation, the capacity to transform into other cell types with specific functions. Self-renewal is initiated when a stem cell receives a specific set of signals that prompt it to divide. These signals are often biochemical in nature and come from within the cell itself or its immediate environment.

The internal mechanisms that govern this process hinge on the regulation of certain genes that maintain the cell's undifferentiated state. Proteins known as transcription factors play a crucial role here. They either promote the expression of genes necessary for self-renewal or suppress those that would lead the cell to differentiate. An example of such a transcription factor is Oct4, which is vital for maintaining the pluripotent state of embryonic stem cells.

Externally, a stem cell's surroundings, or 'niche', provide growth factors and extracellular matrix components that signal the cell whether to divide or begin the process of specialisation. An example is the bone morphogenetic protein (BMP) family, which can promote bone cell differentiation.

Differentiation occurs when the fine balance between self-renewal signals and differentiation cues shifts towards the latter. This change involves the silencing of self-renewal genes and the activation of genes necessary for the development of specialized cells. Epigenetic modifications, such as DNA methylation and histone acetylation, remodel the DNA structure to either expose or hide genes from the cellular machinery, effectively turning them 'on' or 'off'.

Embryonic stem cells represent a type of cell with high potency, meaning they can generate any cell type. In therapy, they may one day be used to create cells to replace those lost to injury or disease, such as dopamine-producing neurons in Parkinson's disease. Adult stem cells, while more limited in potency, are significant in ongoing tissue repair and maintenance. Induced pluripotent stem cells, on the other hand, are adult cells genetically reprogrammed to exhibit characteristics similar to embryonic stem cells. They offer the potential for patient-specific therapies, minimizing the risk of immune rejection.

The potential applications of stem cells in medicine are vast, from generating healthy tissue for transplantation to providing platforms for drug testing. The ethical considerations involve the source of embryonic stem cells and the moral status of the embryo. The field continues to evolve with a strong focus on understanding these complex ethical concerns alongside the scientific advancements.

Stem cells are primarily categorized into three types, each with distinct capabilities and functions. Embryonic stem cells, found in early-stage embryos, have the most versatility—their function is to provide the initial cellular building blocks that can develop into any cell type of the body, a trait known as pluripotency. Think of these cells as the raw materials in a factory that can be programmed to create any product needed.

Adult stem cells, also known as tissue-specific or somatic stem cells, exist throughout the body after development. They function as a repair system, maintaining and repairing the tissue in which they are found. Imagine adult stem cells as maintenance teams stationed in different departments of the factory with the skills specific to that department's needs.

Induced pluripotent stem cells (iPSCs) are adult cells that scientists have engineered to revert to a state similar to embryonic stem cells. Consequently, iPSCs acquire the ability to develop into any cell type, which means that cells from a patient's skin, for example, could be reprogrammed and then used to generate cells for treatment, like nerve cells or heart cells. This technology paves the way for personalized medicine, where treatments are tailored to the individual.

Understanding these different stem cell categories helps illuminate their respective roles in development, tissue maintenance, and therapeutic potential, highlighting both the versatility and the specialized nature of these unique cells. Each category's functions are critical, with embryonic stem cells laying the foundation for all cell types, adult stem cells contributing to ongoing repair and turnover, and iPSCs opening new avenues for customized medical treatments.

Let's take a deeper look at the remarkable world of embryonic stem cells, often likened to a master class of all-rounders capable of becoming anyone in the future. These cells possess something called pluripotency, much like a universal key that can unlock any door to the many rooms in the mansion of specialized cells within the body. This key is safeguarded by transcription factors, essentially the security guards of cellular identity, with names like Oct4, Nanog, and Sox2. These factors patrol the genetic blueprint, ensuring that the cells keep their all-access pass.

When it comes time for a cell to settle down and specialize, it's like picking a single room to call home, abandoning the roam of the mansion. Signals from outside the cell, like those coming from growth factors or neighboring cells, act like real estate agents, guiding the cell to its new 'home' specialization. Imagine the walls of that room gradually getting adorned with specific decor—representing epigenetic modifications—that determine the function of that room; this is the cell's final transition from pluripotency to a specialized identity.

Adult stem cells are more like local handymen, already specialized with a toolkit suited for particular jobs in their local neighborhood, like fixing leaks in the plumbing or electrical issues similar to repairing tissue. They receive signals, or work orders, that tell them where to go and what to repair, ensuring the neighborhood remains in top-notch condition.

Now, think of induced pluripotent stem cells as the result of a complete career change, where an adult, say an accountant, goes back to school to relearn and become a jack-of-all-trades again. Scientists achieve this by delivering specific factors, like Oct4 and Sox2, into mature cells. They use vectors, akin to school applications, to re-educate the cell to forget its accounting job and to be ready for any role again—an amazing possibility for tailor-made treatments where the body's own cells can be engineered to heal itself.

In the practical world, each of these stem cell types holds immense potential. Embryonic stem cells serve as the go-to blueprints for constructing any cell type, while adult stem cells are the on-the-spot fixers for wear and tear. iPSCs offer a chance at a cellular second youth, potentially revolutionizing personalized medicine. However, with such power comes debate, as the ethics of using embryonic stem cells or the safety of reprogramming cells weigh heavily in scientific and public discourse. As we venture further into understanding these incredible cells, we must balance our capabilities with careful consideration of the implications.

Stem cells are essential to human health, fulfilling roles in growth, development, and healing that nothing else in the body can accomplish. During development, stem cells function much like the construction workers and raw materials needed to build the intricate structure of the human body. They specialize and mature into the various cell types that form our organs and tissues, laying the groundwork for all bodily functions.

As we grow and age, these cells continue to play a critical role, acting as a repair mechanism for the body. Like a dedicated maintenance crew, adult stem cells replace damaged or worn-out cells, ensuring that tissues can recover from injury and the normal wear and tear of daily life. This cellular turnover is vital in maintaining the body's integrity and functionality.

When injuries occur, such as a cut on the skin or damage to a muscle, stem cells at the injury site multiply and transform into the specific cells necessary to repair the damage. This process is similar to a rapid response team that swoops in to restore normalcy after an emergency.

By understanding the fundamental role stem cells play, we gain insight into how the body grows, repairs itself, and maintains overall health. Recognizing the importance of each detail surrounding stem cells—from their unique ability to differentiate to their critical repair functions—adds depth to our appreciation of the complexities of human biology and the potential innovations in medical treatment. This knowledge is not only integral to scientific advancement but also empowers us with a better understanding of our own health and wellbeing.

Let's take a deeper look at the fascinating transformation of a stem cell into the diverse cast of cells that make up our body. Imagine a stem cell as an actor on the stage of life, equipped with a script full of potential roles but waiting for the director—the body's signaling molecules—to call out the part it will play.

Initially, our versatile actor — the stem cell — stands ready within the wings, undifferentiated, and holding a script filled with blank pages. These pages are the cell's genes, currently unmarked by destiny. Then comes the cue: signaling molecules from the surrounding environment, working like stage cues and whispers from the director, begin to fill these pages, scripting whether the cell will emerge as a sturdy muscle cell, a swift nerve cell, or any other character in the bodily ensemble.

As the signals accumulate, they flip genetic switches, prompting the cell to read certain lines and ignore others. This process unfolds through the meticulous work of transcription factors, which are like the highlighters and red pens that mark key lines in the script for our actor to study. Each highlighted gene corresponds to a protein to be produced, gradually outfitting the cell with the right costume and props for its assigned role.

The adult stem cells, akin to understudies, come into play when the lead performers — the body's primary cells — fall to injury. They heed the distress signals, acting as rapid response teams that leap into action, dividing and transforming to replace the lost cells and restore the tissue's integrity. They receive the baton, or signal, from damaged cells and race into the fray, bandaging wounds and mending breaks.

Through understanding these minute details, we appreciate not only the stem cell's versatility but also its disciplined response to the body's call for repair and renewal. It's a process as orderly and dynamic as a well-directed play, where every actor knows their part and comes together to create the grand spectacle of life.

Imagine you're at a career fair, standing at the crossroads of countless professional paths, each promising a different future. Now, picture a stem cell in much the same scenario. Just as a student might begin with a broad education, a stem cell starts with a comprehensive set of instructions that could lead to any number of outcomes. As time goes by, specific influences — be they mentor advice for the student or chemical signals for the cell — nudge the direction towards specialization.

For the student, this might mean picking a major in university, just as a stem cell 'decides' its function—a process known as differentiation. Let's say our student chooses medicine; they're now on track to become a doctor but still have a way to go. Similarly, the stem cell, now destined to be a heart cell, needs to undergo training, taking on unique characteristics and responsibilities required of cardiac tissue.

The student, now a medical resident, fine-tunes their skills through highly specialized rotations, akin to the stem cell acquiring all the protein tools it needs to perform its heart cell duties. Eventually, there's a graduation — our student becomes a full-fledged surgeon, and our stem cell becomes an integral part of the beating heart.

This journey from 'jack of all trades' to 'master of one' is the essence of cellular specialization, an essential and intricate process that makes our bodies the complex systems they are. Every cell, like every professional, has a role that contributes to the greater whole. And understanding this is like recognizing the diverse and crucial range of jobs that keep our communities thriving.

Here is the detailed breakdown of the stem cell differentiation process, explained through the lens of analogies to illuminate each intricate step:

- Triggering Factors:
    - Think of transcription factors like Oct4 and Sox2 as skilled talent scouts. They identify stem cells with the potential for a particular role, marking them for specific functions much like a talent scout spots an actor for a part in a play.
    - The stem cell does not exist in isolation. Its interactions with surrounding cells and the extracellular matrix can be compared to a budding artist interacting with their peers and mentors, shaping the direction of their growth and specialization.
    - Reception of biochemical signals, such as cytokines and growth factors, is akin to the artist receiving an invitation to an audition that could define their career. These signals dictate whether the stem cell will commit to a particular path.

- Genetic Regulation and Signaling Pathways:
    - Like a script determining an actor's lines, certain genes are activated or inactivated, leading to the stem cell assuming its specialized role.
    - Enhancers and silencers can be thought of as directors and editors, deciding which genes will be expressed prominently and which will be muted in the cell's final performance.
    - Pathways like Wnt, Notch, and Hedgehog serve as the genres or schools of acting – be it drama, comedy, or action. They guide the overarching theme of the cell's development and eventual specialization.

- Stages of Cell Specialization:
  - The initial cell division and expansion are the auditions and callbacks — a series of screenings where the stem cell proves its versatility and potential before getting the part.
  - Transitional phase: The cell starts rehearsing its role, no longer just auditioning but perfecting the nuances of the character it will become.
  - Final commitment occurs when the stem cell signs the contract, firmly taking on the role of a particular cell type, like a cardiomyocyte, akin to an actor finally stepping on stage for opening night.
  - The acquisition of unique functions and structures can be thought of as the actor's costume and props coming together, fully embodying the role so that the performance can contribute to the story — or, in this case, the organ's function.

- Tissue Regeneration:
  - Adult stem cells recognize tissue damage as community helpers would notice areas that need renovation. They are the local contractors who know precisely how to mend the flaws and restore the community's structure.
  - The sequence of events that lead to repairing the damaged tissue: an emergency call is sent, the responders gather their tools, and they get to work reconstructing and revitalizing the affected area.
  - Inflammation and remodeling are like the clean-up crew and the refurbishing team that not only clean up the aftermath of a disaster but also renew and upgrade the infrastructure.

Through these analogies, each stage of the stem cell differentiation process is demystified, shown to not just function mechanically but to matter deeply in the living narrative of the body's health and repair. Understanding this process invites us to admire the body's inherent wisdom and the potential for stem cells to revolutionize treatment and healing.

Imagine stem cells as a comprehensive, high-end toolkit, complete with gadgets and instruments that a world-class surgeon might use. Each tool in the kit has a specific function, designed to address and repair a range of health issues. Just as a scalpel precisely cuts through tissue or a clamp stops a bleed, stem cells have the remarkable ability to become the exact type of cells necessary for healing our bodies. From mending a broken bone by transforming into bone cells, to replacing damaged heart muscle post-heart attack, these cells are the versatile tools in the body's natural repair kit. Each "tool" works in harmony, orchestrated by the body's own intelligence, to heal from injuries, fight diseases, and restore functionality. This cellular toolkit is not just a collection of instruments for repair; it represents the future of personalized medicine, where treatments are as tailored and precise as the finest surgeon's equipment, offering new hope where conventional therapies might fall short.

Let's take a deeper look at the body's remarkable ability to signal its own repair through stem cells, explaining it in a way that sits right at home with our day-to-day lives. Think of stem cells as the ultimate handymen on call, waiting for a signal to rush to the aid of damaged tissues. When we injure ourselves, the body sends out an SOS—biochemical signals like cytokines and growth factors that act like a 911 call to these cellular repairmen.

These signals are the homing beacons, the flashing lights and the siren calls that direct the stem cells to the exact location of the injury, much like GPS and emergency lights help an ambulance find the quickest route to the crisis site. Once the stem cells reach the damaged area, there's a huddle, a team meeting where cells at the injury site share information about the damage through complex signal exchanges; this is the damage assessment phase.

From there, the stem cells, now well-informed repairmen, start their work. They get their tools ready, which in this case are proteins specific to the task at hand. This tool preparation is driven by the cell's inner machinery responding to the genetic blueprint, which gets edited according to the need—imagine a handymen's tool belt being customized on the spot for particular repairs.

Gene expression, which decides which tool comes out of the belt, is the foreman's instructions. Specific genes are turned on or off, in essence deciding whether a stem cell will become a brick (bone cell) or a new pipe (blood vessel) to replace the broken parts. It is a meticulous process, as each tool or protein must be the perfect fit for the repair job, demonstrating the biological precision our bodies inherently possess.

Appreciating this deep layer of intelligence our bodies carry subtly reminds us of the complexity yet the artistry with which life operates within us, turning even the tiniest cellular interactions into life-sustaining actions. Understanding this, we not only marvel at what the body can do but also discover a new respect for the science of healing and the therapeutic potential that stem cells present. It's not just a response to injury; it's a testament to the body's innate resilience and foresight.

In the realm of scientific discovery, Shinya Yamanaka stands tall as a trailblazer who carved a new path in the understanding of stem cells, much like the intrepid Marco Polo opening new routes along the Silk Road. Just as explorers of old set sail to discover unseen lands and chart the unknown, Yamanaka embarked on a quest into the cellular cosmos to unlock the potential of stem cells. His groundbreaking work on induced pluripotent stem cells (iPSCs) was a feat akin to turning lead into gold, reshaping adult cells back into their youthful, versatile state. This monumental achievement not only challenged our fundamental notions of cellular development but also opened up a treasure trove of possibilities for medical advancements, much like the unveiling of a hidden world map marked with X's denoting treasure troves of knowledge and cures. The dedication and innovative spirit of pioneers like Yamanaka underscore the relentless human pursuit of knowledge, akin to scaling the highest peaks or reaching the depths of the ocean—always pushing the boundaries of what is possible.

Let's take a deeper look at Shinya Yamanaka's groundbreaking efforts, as revolutionary as an alchemist turning base metals into gold, reprogramming adult cells to revert to a state indistinguishable from embryonic stem cells. This process begins akin to a software update on a computer, with specialized vector delivery systems introducing a new set of instructions to an adult cell. These vectors are like courier services, delivering the Yamanaka factors—Oct4, Sox2, Klf4, and c-Myc—key codes which reprogram the cell's DNA, effectively wiping its functional memory and returning it to a pluripotent, embryonic-like state.

Imagine taking a seasoned detective and sending them back to the police academy to relearn everything from scratch; this is the journey Yamanaka factors initiate. The introduction of these factors prompts extensive epigenetic remodeling, where the cell's histones and DNA
are chemically modified, effectively erasing signs of aging and learned cellular behavior, rejuvenating the cell completely.

Yamanaka faced trials similar to those a pioneer in a new world would, with ensuring the reprogrammed cells were safe for use and retained their stability over time. Through experimentation, akin to an artist perfecting their technique, he refined the methods to improve both the efficiency and safety of iPSC generation.

The advent of iPSC technology, comparable to the invention of the printing press, has revolutionized the landscape of regenerative medicine, drug development, and disease modeling. It has empowered scientists to customize treatments, develop more accurate disease models, and open up a new frontier in personalized medicine.

Nonetheless, these discoveries carry ethical burdens akin to the responsibilities borne by inventors who must consider the wider impact of their creations. The ethical debates surrounding iPSCs touch upon the concerns of genetic manipulation and the potential for unforeseen consequences, leading the scientific community into a discourse as intricate as the cells themselves. This conversation is essential to navigate as we harness the immense potential of iPSCs to heal and understand the human body, ensuring progress in harmony with ethical standards.

Embarking on the exploration of "The Stem Cell Universe" reveals the profound adaptability and infinite promise embedded within stem cells. These biological entities stand out for their distinctive capacity to differentiate into a diverse array of specialized cells, playing critical roles in growth, repair, and regeneration throughout the human body. As we unravel the complexities of cellular differentiation, and regeneration, the vast potential of these cells begins to crystallize—notably in their applications for innovative medical therapies and research. Stem cells have the unique ability to respond to the body's needs, highlighting their promising future in potentially treating a range of diseases and conditions. It is this remarkable versatility that places stem cells at the forefront of medical science, opening new horizons for healing and understanding human health.

CHAPTER 2 THE LIFECYCLE OF A STEM CELL

Get ready folks, we will now dive into the lifecycle of a stem cell—these remarkable units of life with the extraordinary ability to create the diverse array of specialized cells that make up our bodies. Just as a seed grows into a tree through various developmental stages, a stem cell also undergoes a series of transformations, each guided by a biological process that is both complex and elegant. These cells are essential to the development and maintenance of tissues and organs, possessing the incredible potential to repair and regenerate, impacting health and disease treatment profoundly. In this chapter, we explore the journey of stem cells from their origin to their mature forms, examining their vital roles and the impact they have on the vast landscape of human biology. Join us as we unravel the science behind the potential of stem cells, setting a foundation for a deeper understanding of their pivotal place in life's tapestry.

Stem cell development proceeds through rigorously controlled stages, much like a student's progression from a general curriculum to a focused specialty. Initially, stem cells exist in a state called 'pluripotent', which means they have the potential to turn into almost any cell type in the body – akin to an apprentice with the capability to master any trade. This potential is analogous to having a blank canvas, providing a foundation for various futures.

The first transition in stem cell development is from this open-ended pluripotent state to a more focused set of possibilities, known as lineage commitment. Here, the cell is similar to a student declaring a major, narrowing down the potential career paths. The stem cell, now lineage-committed, specializes further. For instance, it might become a blood cell, a skin cell, or a neuron, each type fulfilling a unique and necessary function within the body's complex ecosystem.

Think of this specialization process as choosing the tools for a specific job – the blood cell carries oxygen like a delivery truck, whereas the neuron sends messages like a phone line. This stage of development emphasizes a cell's transition from being a jack-of-all-trades to mastering the specific skills required for its role in the body.

Each step in this process is guided by internal and external signals that act like mentors and real-world experiences shaping the student's path. In the end, the mature cell integrates into the existing systems of the body – joining a team of cells in harmonious function. By understanding these developmental stages in detail, we can appreciate the intricate choreography behind the body's function and the potential for stem cells in medical advancements, offering promises of repair and renewal where it was once thought impossible.

Let's take a deeper look at the intricate web of signals that guide a stem cell through its life journey. Imagine a stem cell like a young student entering a university, faced with an array of career paths. Just as advisors and course selections steer the student, a network of signaling pathways directs the stem cell towards its destiny.

As the student attends classes—some mandatory, others by choice—their academic career begins to take shape. In the cellular world, this is mirrored by internal chemical signals acting on the stem cell, triggering certain genes to switch on or off. It's like the student declaring a major, setting the course towards becoming an engineer, a biologist, or an artist. These pivotal points in a stem cell's life are mediated by transcription factors—the academic advisors of cell fate—counseling which genetic programs to follow, ultimately committing it to a specific function.

With each decision, the student's expertise grows, and the stem cell's characteristics define further, preparing for a role in the greater biological society. Like mastering the skills required for a profession, the stem cell specializes, honing the tools—proteins—needed for its specific job within the body. This is its graduation moment, where it's no longer a generalist but a specialist contributing to the body's well-being.

After graduation, the once-student integrates into the workforce; similarly, the now-specialized cell finds its niche in the body. It becomes part of a community—like a heart, liver, or brain—where teamwork is key. This is an orchestra of cells where each must harmonize with the others for optimal performance.

Understanding these stages unveils the complexity and beauty of life at its cellular foundations. With each analogy, from a student's choices to a symphony orchestra, we translate the biological intricacy of stem cell differentiation into everyday experiences, shedding light on a process that is as fundamental as it is astounding.

Just as the health of a seedling hinges on the quality of the soil and the climate it takes root in, a stem cell's development is profoundly influenced by its surroundings. This environment, known as the stem cell niche, serves as the context in which these potent cells reside and receive cues that can determine their fate and function. Signal molecules released within the niche act like the nutrients and water absorbed by a seedling, instructing the stem cell to either maintain its undifferentiated state or begin the process of specialization, evolving into the specific cell types needed for action and repair.

Temperature and weather patterns guide a plant's growth cycle—similarly, factors such as blood supply, physical contact with other cells, and the presence of certain chemicals make up the 'climate' of a stem cell niche, shaping each stem cell's destiny. Just as two plants, side by side, might grow differently due to microclimate variations, adjacent stem cells might develop along divergent paths because of slight differences in their immediate environments.

This dynamic interplay of external conditions and intrinsic cellular response orchestrates the harmony of our body's cellular composition. The careful balance managed in the stem cell niche ensures the body's continuous regeneration and repair, akin to a garden that needs the right balance of sunlight and shade, water and airspace to yield a rich variety of plants and flowers. Understanding how these extrinsic factors govern stem cell behavior is key not just to comprehending their nature but also to harnessing their capabilities for advancing medical treatments.

Here is the detailed breakdown of the factors within a stem cell niche that influence stem cell differentiation:

- Key signal molecules:
  - Growth factors: Imagine these as personal trainers in a gym, they direct the stem cells' workout regimen, conditioning them to become the specific muscle (cell type) needed for the body's fitness (function).
    - They bind to receptors on stem cells, initiating a cascade of intracellular events that guide cell fate.

- Cytokines: Consider these the local news reporters, broadcasting the status of the body's health and thus influencing the stem cells' decisions, much like how news might influence your daily activities.
- They can stimulate or inhibit stem cell growth and are critical in inflammatory responses.
- Hormones: Like career advisors, they provide long-distance guidance and can alter a stem cell's trajectory over time, impacting their maturation and function.
- Hormonal signals often coordinate stem cell activities with the body's overall state, ensuring a harmonized response to physiological needs.

- Cellular communication:
- Direct contact with neighboring cells: Think of this as a team huddle, where cells share their game plans through physical contact, influencing each other's next moves.
- Cell-cell interactions can result in the direct transfer of signals that specify the stem cell's role, akin to teammates passing along instructions.
- Matrix components: Similar to an architect's blueprint, the structural elements of a cell's surroundings provide critical information for shaping its development.
- The extracellular matrix's composition and rigidity can inform the stem cell whether it's destined for a role like foundation (bone) or cushioning (fat).

- Physical properties:
- Oxygen levels: Picture a scuba diver adjusting to different depths. Low oxygen might signal the niche is a niche 'deep underwater', guiding the stem cell towards becoming an efficient oxygen user like a red blood cell.
- Hypoxic conditions can promote the retention of a stem cell's pluripotency or drive specialization, depending on the context.
- Rigidity and topography: This is the stem cell's 'terrain'. A soft, flat surface might guide it into becoming something like skin, while a rigid, rough surface might indicate a future as bone.
- Variations in mechanical properties of the niche can direct stem cells toward specific lineages by modulating the cell's internal scaffolding and tension.

This interplay within the niche establishes a unique 'microenvironment' for each stem cell, guiding its specialization as surely as a climate shapes its local flora. By understanding these intricate details, we open a window into the mastery behind bodily functions and the promise of harnessing this knowledge in medicine.

Bone marrow transplantation is a medical procedure that can be lifesaving for individuals with conditions like leukemia. This process begins with the extraction of healthy stem cells from a donor's bone marrow – the soft, spongy tissue inside bones where blood cells are produced. These cells are the foundation for various blood components, including the white blood cells crucial for immune function, red blood cells that carry oxygen, and platelets that aid in clotting.

The recipient's diseased marrow is then eradicated, often through chemotherapy or radiation, creating a clean slate for the new stem cells to engraft. The healthy donor stem cells are infused into the recipient's bloodstream, much like a blood transfusion. These cells navigate their way to the recipient's bone marrow cavities, where they begin to settle and multiply.

Over time, these transplanted cells restore the bone marrow's ability to produce healthy blood cells, effectively repopulating it with functional immune cells that can combat leukemia, red blood cells, and platelets. It's a process akin to repopulating a deforested area with new saplings, where the new trees eventually grow and re-establish the forest ecosystem.

The significance of bone marrow transplantation lies not only in its ability to replace diseased cells with healthy ones but also in the potential to provide a new, functional immune system capable of fighting off cancer cells. It's a clear example of stem cells' therapeutic potential and a testament to their crowning property: the ability to give rise to various cell types that our bodies need for recovery and health. This procedure, while carrying risks and complications, illuminates the profound capabilities of stem cells in medical treatment, offering hope and a renewed chance at life for many patients.

Bone marrow transplantation begins with a match. Doctors test the patient and potential donors for human leukocyte antigens (HLA), essential proteins on the surface of cells that the immune system uses to recognize which cells belong in the body and which do not. Think of HLA as a cellular ID card; a close match between patient and donor can reduce the risk of transplant complications.

Once a donor is matched, the process of stem cell extraction starts. For the donor, it involves either a surgical procedure to take cells directly from bone marrow or a non-surgical method where blood is drawn, stem cells are separated, and the remaining blood returned to the donor.

With the stem cells harvested, attention turns to the recipient, whose existing marrow must be wiped clean, usually through chemotherapy or radiation. This step ensures that the new stem cells have space to grow and that the underlying disease is treated.

The healthy stem cells are then infused into the patient's bloodstream, similar to a blood transfusion. After the infusion, the cells make their way to bone marrow cavities, where they begin to engraft – or take root – and start producing healthy blood cells.

But the process doesn't end with engraftment. Recovery from bone marrow transplant involves carefully monitoring for any signs of graft-versus-host disease (GVHD), where the new immune cells attack the patient's body. Management of GVHD typically includes medication to suppress the immune response while the transplanted cells settle into their new environment.

For weeks to months following the transplant, doctors closely watch the patient, ensuring that the new marrow is functioning correctly and creating blood cells without issue. Immune system strength is evaluated regularly, and precautions are taken to protect the patient from infections during this vulnerable period.

Through careful preparation, meticulous procedure, and attentive post-care, bone marrow transplantation harnesses the power of healthy stem cells to renew a patient's blood and immune system. This delicate process is a marvel of modern medicine, offering new chances at life for those facing otherwise lethal diseases.

In the sweeping expanse of cellular biology, the work of Ernest McCulloch and James Till stands out like the first map of a previously uncharted land. These two scientists, through diligent research and experimentation, identified what we now know as stem cells—the versatile cells in the bone marrow responsible for generating blood cells. Their pioneering experiments in the early 1960s demonstrated that these cells could self-renew and differentiate into various blood cells, much like a single seed can grow into a whole tree, complete with branches and leaves of diverse cells.

Their discovery was akin to finding the motherlode in a mine; not only did they locate the cells, but they also developed the methods to prove their unique properties. This scientific endeavor set the stage for all future stem cell research, opening doors to therapies for diseases once deemed incurable. Thanks to their work, we've seen successes in treating conditions like leukemia with bone marrow transplants—a real-world application of their foundational research.

The insights provided by McCulloch and Till continue to propel discoveries in the vast realm of cell biology, enhancing our understanding of regeneration and repair. As trailblazers, they marked the trails for others to follow, leaving a lasting legacy that continues to guide the scientific community as it navigates the dense 'forest' of cellular biology. Their contributions have not only furthered scientific knowledge but have also saved countless lives, demonstrating the power of curiosity and the enduring value of charting the unknown.

Let's take a deeper look at the groundbreaking work of Ernest McCulloch and James Till, which laid the foundation of stem cell science as we know it. Back in the 1960s, they started with a hunch, much like detectives on the trail of a mystery – their clue was that bone marrow must contain some kind of master cell that could replenish all blood cells.

To prove their theory, they conducted experiments that now seem as fundamental as the alphabet is to language. They injected bone marrow cells from one mouse into another whose bone marrow cells had been destroyed by radiation. What unfolded was akin to planting seeds in a barren garden and watching new growth take root. The transplanted bone marrow cells not only survived but multiplied, and the once empty 'garden' of the recipient mouse's marrow bloomed with new blood cells.

Further experiments revealed nodules on the spleens of recipient mice, each a direct result of a single transplanted marrow cell – these nodules were concrete, visible evidence of individual stem cells growing and creating colonies. Like a single blueprint giving rise to different structures, these cells could differentiate into a variety of blood cells, revealing their extraordinary potential.

The scientific community initially met their findings with skepticism, as any radical idea might be. Yet, over time, the reliability and implications of their work became clear, setting off an avalanche of research and understanding. Bone marrow transplantation, once a distant dream, became a reality, with their research paving the way for treating blood cancers like leukemia.

Today, we see the impact of McCulloch and Till's experiments in every stem cell therapy and each new understanding about how our bodies renew and repair themselves. Their work, once a single point of light in an obscure landscape, has grown into a beacon that guides ongoing exploration in the vast 'forest' of cellular biology. Their legacy is not just in the knowledge they uncovered, but in every life that's been touched by the medical advancements their discoveries made possible.

In the realm of stem cell research, ethical considerations are critically important and can be likened to the management of a forest. Just as the conservation of a forest involves careful decisions about which resources to use and which to preserve, stem cell research requires a measured approach to how we utilize and respect these powerful biological resources. One of the key ethical issues in this field is the source of stem cells, particularly those derived from human embryos. This raises questions similar to those about preserving ancient trees; while there may be great potential in the research, there's also a call for profound respect for the origins of these cells.

Furthermore, the manipulation of stem cells entails another layer of ethical scrutiny. As we modify such fundamental components of life, questions arise about the natural order—parallels can be drawn to introducing non-native species into an ecosystem, which could offer benefits but also pose risks. These important considerations compel researchers to weigh the potential gains against the moral and societal implications.

Pivotal to this discussion is informed consent, where donors must be fully aware of how their cells will be used, mirroring the clarity required when managing a forest's resources for the good of many. The overarching concept here is one of balance—advancing scientific understanding and medical progress, while ensuring that ethical lines are both respected and clearly drawn. As we tread forward in the exploration of stem cells, maintaining this equilibrium will remain vital to the integrity and sustainability of both the science and its impact on society.

Step 1: Identify Ethical Issues

Start by understanding the origins of stem cells. Embryonic stem cells, derived from human embryos, raise questions about the beginning of human life and rights. This is a crucial ethical issue that requires careful consideration, similar to evaluating the impact of felling ancient trees in a forest for research or therapeutic use.

Step 2: Risk Assessment

Evaluate the benefits and risks associated with manipulating stem cells. Potential advantages include the development of new therapies for chronic diseases, but risks may involve unforeseen long-term effects. Compare this to introducing a non-native species in a forest—while it may have short-term benefits, the long-term impact on the ecosystem can be hard to predict and potentially irreversible.

Step 3: Informed Consent Implementation

Create a transparent and robust process for obtaining informed consent from donors of stem cells. This involves making sure donors are aware of how their cells will be used, the potential benefits, and associated risks, ensuring their participation is voluntary and well-informed. It's akin to managing a forest's resources responsibly, where all parties affected by the decision-making are considered and informed.

Step 4: Maintain Ethical Balance

Develop strategies to maintain ethical balance amidst advancing research. This includes strict adherence to regulations, ongoing ethical review, and open dialogue with the public—much like a forest management plan that balances economic benefits with conservation efforts. It's essential to remember that while scientific progress is valuable, it should not compromise ethical standards.

Looking ahead, the potential for stem cell technology in medicine is broad and promising. These cells, with their ability to develop into a vast array of human tissue types, hold the key to treating a multitude of conditions. We're looking at a future where diabetes patients may receive pancreatic cells to manage their insulin levels accurately, or people with heart disease might have damaged tissue repaired with healthy, beating cardiac cells.

As research progresses, we may see stem cells used to grow entire organs for transplant, bypassing long waiting lists and the issue of organ rejection. This could revolutionize organ transplantation, making it accessible and safe for more patients. On a smaller scale, but no less revolutionary, stem cells could be used to heal wounds or burns without scarring, remodel damaged joints, or even reverse the effects of degenerative diseases such as Alzheimer's.

These advancements are not science fiction but are based on solid, ongoing research that is uncovering more about stem cell capabilities each day. In the not-too-distant future, these 'seeds' of cellular technology could sprout into a 'forest' of new treatments, each branch offering hope for a different medical challenge. The journey from laboratory to clinic is complex and takes time, but the path is being forged with careful and methodical science. This vision of the future is not without its hurdles, but the promise it holds for improved health and longevity is profound and worth pursuing with diligence and hope.

Let's take a deeper look at stem cell research as it stands today, particularly in the exciting field of regenerative medicine. Picture stem cells as master key cells with the potential to unlock the development of any specialized cell type needed to heal the body. Scientists have been developing techniques to coax these master keys into becoming specific cells, like training an apprentice into a master craftsperson.

For example, researchers are working on turning stem cells into pancreatic beta cells, the insulin producers that are in short supply in diabetes. It's similar to teaching someone the intricate art of watchmaking to repair a complex timepiece. Cardiac cells for heart disease treatment are another goal. This involves guiding a stem cell down the path to becoming a thriving muscle cell, pumping with the same rhythmic precision as a drummer in a band.

The ambition goes even further with the possibility of growing whole organs from stem cells. This is like using a single brick to not just patch a wall, but to build an entire house. The progress here has been substantial, with early successes in creating organ-like structures known as organoids. However, scaling this up to full organs is an enormous challenge, akin to the transition from crafting model cars to manufacturing a fleet of actual vehicles.

One major hurdle to transfusing or transplanting these crafted tissues or organs is immune rejection, a problem comparable to a security system that cannot recognize its new, legitimate user. To overcome this, scientists are exploring ways to make stem cells 'invisible' to the immune system or genetically matching them to the recipient, much like creating a key tailored to a specific lock.

Of course, ethical concerns also loom over this field, inviting as much debate as the discussions over land conservation or genetic modification in crops. It is crucial to navigate these issues with transparency and integrity, securing informed consent from cell donors, and ensuring equitable access to these burgeoning therapies.

The journey of stem cells from the lab bench to clinical application is intricate, but each step forward is made with the goal of improving human health. Advancements in the lab translate to hope in clinics and hospitals—a hope for a future where damaged tissues and organs are not an irreversible loss but a repairable reality.

In conclusion, the lifecycle of a stem cell captures the remarkable journey of transformation and adaptability. From their origin in the marrow – an embryonic stage akin to the seed of a tree – these cells branch out into a variety of forms, each fulfilling a specific function in the body's ecosystem. Their critical roles encompass growth, healing, and regeneration, akin to a tree that supports life with its oxygen, shelter, and sustenance. Throughout this chapter, we've traced the path of stem cells from their pluripotent beginnings to their specialized endpoints, mirroring the growth and development of a tree in diverse environments. This analogy reflects both the promise of stem cells in medical advancements and their inherent versatility, characteristic of a resource that holds the potential to revolutionize our approach to health and disease.

CHAPTER 3 THE BLUEPRINT OF LIFE

You've made it to Chapter 3, "The Blueprint of Life". In this chapter, we explore DNA, the molecule that serves as the master blueprint for all living organisms. DNA encodes the genetic instructions vital for the functioning and reproduction of every cell. It is the basis for cellular structure, function, and diversity, carrying the plans necessary to build the vast range of cells—from those that form the structure of our bodies to those that govern our immune responses. Understanding DNA is paramount, as it not only dictates the unique characteristics of an individual but also has the power to provide insights into a multitude of genetic disorders. Just as a building requires a blueprint to take shape, every cell relies on DNA to guide its formation and role within the larger biological system. As we delve into the pages ahead, expect to unravel the detailed workings of DNA and gain a clearer understanding of its central importance in life's intricate architecture.

Genetic coding is a fundamental molecular process that shapes every living creature, guiding stem cells as they develop into the multitude of specialized cells that form an organism. At the core of this process is DNA, a long, coiled molecule that resides in the nucleus of every cell. It comprises four basic building blocks: adenine, thymine, cytosine, and guanine. Think of these building blocks as the letters in a complex instruction manual that dictates everything a cell must do.

During cell division, DNA unwinds and splits down the middle, allowing each strand to serve as a template for duplicating the genetic code—a process reminiscent of taking a photocopy of an essential document. This ensures that every new cell inherits the exact same set of instructions as its parent, which is critical for maintaining the body's functions.

When a stem cell receives signals to specialize, select segments of DNA, known as genes, are read and translated into RNA—similar to how an architect's blueprint is translated into a working model. The RNA then exits the nucleus and heads to a cell's ribosomes, which are akin to tiny factories. Here, the RNA's instructions are used to assemble proteins by linking together amino acids in precise sequences. These proteins become the building blocks of cells and tissues, driving the development and maintenance of the body.

The intricacy of differentiating a stem cell into a specific type, such as a muscle cell or nerve cell, hinges on which genes are expressed, or 'turned on,' and which are 'silenced.' This selective gene expression is key to achieving the vast diversity of cells within an organism, shaping the intricate and dynamic system that constitutes life.

Understanding genetic coding is not just an academic pursuit but has practical implications for medicine and biotechnology. For example, identifying how genetic codes contribute to disorders can lead to targeted therapies. By clearly explaining these concepts, we unravel the elegant complexity behind organism development and appreciate the pivotal role of genetic coding in life's rich tapestry.

Transcription and translation are two key processes in the central dogma of molecular biology, transforming the code within DNA into the proteins that perform almost every function in our cells. Here's a step-by-step guide to these processes, with a focus on how they enable stem cells to become specialized cells within the body.

Step 1: Initiation of Transcription
Transcription begins when proteins called transcription factors bind to a specific region of DNA called the promoter. The promoter acts like a launchpad, signaling the start of a gene. It's comparable to the foreman at a construction site deciding where to build.

Step 2: Assembly of the Transcription Complex
After the transcription factors are in place, an enzyme known as RNA polymerase attaches to the DNA at the promoter region. This step is akin to bringing in the main piece of machinery needed to start construction – a powerful crane that will help build the structure.

Step 3: Creation of Messenger RNA (mRNA)
RNA polymerase travels along the DNA, 'reading' the code and creating a complementary strand of RNA, called messenger RNA (mRNA). This is similar to running a credit card along a zipper, with the teeth representing the DNA sequence and the resulting separation symbolizing the mRNA strand.

Step 4: RNA Processing

The newly created mRNA is a rough draft and needs processing. It undergoes splicing, where unnecessary sections (introns) are removed, and the remaining sections (exons) are spliced together. Think of this step as editing a long film into the final cut that will be seen by an audience.

Step 5: mRNA Exits the Nucleus

Once processing is complete, mature mRNA exits the nucleus through nuclear pores. This resembles a finished set of architectural plans leaving the office to go to the job site where the building will rise from the blueprint.

Step 6: Translation and Protein Synthesis

Translation is where the mRNA is 'read' by a cellular structure called the ribosome, which functions much like a factory assembly line. Transfer RNA (tRNA) molecules bring the appropriate amino acids (the building blocks of proteins) to the ribosome, which assembles them in the sequence dictated by the mRNA.

Each three-letter 'word' (codon) on the mRNA corresponds to a specific amino acid, similar to a construction worker reading instructions to determine which piece to add next. The ribosome moves along the mRNA, adding one amino acid at a time until the protein is fully assembled.

In Stem Cell Differentiation:

This entire transcription-translation process is critical for stem cells as they differentiate. Different proteins, produced from different genes being 'turned on' or 'off,' dictate what type of cell a stem cell will become – whether a muscle cell, a nerve cell, or any other type. Like choosing different materials or design features for a construction project, these proteins determine the characteristics and function of the resulting specialized cell.

By precisely orchestrating which genes are expressed and exploring their impact on stem cell behavior, researchers can guide stem cell differentiation for therapeutic purposes, such as repairing damaged tissues or treating genetic disorders. This complex interplay of genetic information is the foundation of cellular and organismal biology, providing exciting possibilities for biomedical advancement.

Imagine your body as an intricate symphony, with each cell playing its unique part, much like musicians in an orchestra. The genetic coding in stem cells is akin to sheet music filled with complex compositions—these are the genes. Each cell in your body has the same vast repertoire, but like a musician choosing to play only certain pieces at a concert, a cell chooses to express specific genes at any given time. This selective gene expression allows stem cells to adapt and become specialized—becoming a heart cell, a brain cell, or any other cell, just as a musician might specialize in the piano, violin, or trumpet.

The process of stem cell specialization is comparable to a professional's journey through various disciplines. Think of a stem cell as a student with a full curriculum, the genetic code being their potential career paths. The signals a stem cell receives act like career advice, leading the cell to specialize, say, into an engineer rather than a painter; the right genes are 'turned on' to equip the student with the necessary skills for their chosen field.

In the grand scheme of the body, the meticulous orchestration of each cell's role ensures that, together, they create a harmonious symphony of life. The beauty of this process lies in the possibility of applying such knowledge to medicine, where we could potentially conduct the symphony in ways that heal and regenerate. Understanding genetic coding and stem cell specialization isn't just about grasping the complex science—it's about appreciating how this biological orchestra plays out in our health and well-being.

Here is the breakdown on the compelling journey of a stem cell as it follows the genetic 'sheet music' to find its place in the body's 'orchestra':

- **The Structure of Genes and Their Regulatory Elements:**
    - Genes: Think of these as individual songs in a musician's setlist, containing specific notes (base pairs) that need to be played.
    - Each gene has coding regions (exons) that directly spell out the protein 'melody' and non-coding regions (introns) that get edited out during the 'soundcheck' (mRNA processing).
    - Regulatory Elements: These are the guidelines for when and how a song is played.
    - Promoters indicate the beginning of a gene, like a conductor signaling the start of a piece, while enhancers can be thought of as amplifiers,

increasing the gene's expression.

- Role of Transcription Factors and Enhancer Regions:
- Transcription Factors: These are the talent scouts that recognize potential hit songs (genes that should be expressed).
- They bind to the promoter region and recruit RNA polymerase, the 'band' that will play the gene's melody.
- Enhancer Regions: Like superfans who boost the band's performance, enhancers increase the likelihood that a gene will be 'heard' across the cell.

- Influence of External Signals on Transcription Factors:
- Signals like hormones or growth factors act as fan requests at a gig, influencing which songs (genes) the talent scouts (transcription factors) choose for the performance.

- Process and Stages of Transcription:
- Initiation is the soundcheck where all the band members (molecular machinery) gather at the promoter stage.
- Elongation is the actual concert, where the RNA polymerase 'band' travels down the DNA, 'singing' the RNA tune.
- Termination is the end of the gene song, where the RNA transcript is complete, and the band takes a break.

- mRNA Processing Includes Capping, Splicing, and Tailing:
- Like adding special effects and cleaning up a track before release, the mRNA receives a protective cap at the start, gets 'edited' to remove introns, and gains a tail at the end to ensure stability and longevity.

- Ribosomes Translate mRNA into Protein:
- Think of ribosomes as DJs, spinning the mRNA track and using tRNA samples to produce a rhythmic chain of amino acids, the 'dance beat' that is a protein.

- Steps When Stem Cells Receive Differentiation Signals:
- Signals prompt stem cells to begin specialization, like an artist choosing a specific genre to master—rocking guitars for rockstars, smooth keys for jazz pianists.

- **Role of Specific Proteins in Guiding Stem Cells:**
 - Master regulatory proteins act as genre specialists, guiding stem cells to make the right career choice, be it joining the percussion section (becoming bone) or taking on vocals (transforming into blood cells).

Each part of this cellular concert is orchestrated precisely, ensuring that stem cells find their perfect role in the body's symphony, contributing uniquely to the harmonious function we recognize as life. Through this lens, we can see just how meticulously life's compositions are arranged, offering methods to tune and perhaps even rewrite our biological songs.

Today's scientific landscape is being remarkably transformed by CRISPR technology, a sophisticated tool allowing researchers to modify genes with unprecedented precision. Originating from a bacterial defense system, CRISPR functions like molecular scissors, capable of locating and cutting specific sequences within the DNA strand. By employing a guide RNA that matches the target sequence, CRISPR can introduce or correct mutations, which can inactivate a faulty gene or repair it.

The precise nature of CRISPR technology has led to significant advancements in understanding and treating genetic disorders. For instance, scientists are working on strategies to use CRISPR to correct the genetic errors responsible for diseases like cystic fibrosis and sickle cell anemia. By correcting these errors, it's as if we're able to fix typos in a crucial document, ensuring that the body's processes run as intended.

However, alongside its considerable potential, CRISPR technology also poses ethical and procedural challenges. Edits to genetic material can have unforeseen consequences, and the possibility of off-target effects—where unintended parts of the DNA are altered—requires careful management. It is a powerful tool, yet one that must be wielded with caution and foresight.

As we advance our capabilities with CRISPR, there is a growing hope for new therapies that can directly address the genetic root causes of previously incurable diseases. This is not merely modifying the genetic narrative of an individual but rewriting the possibilities for the treatment of genetic conditions as a whole. With each careful cut and repair, we step closer to a future where the constraints of genetic ailments might be alleviated or even eliminated.

Let's take a deeper look at the fascinating CRISPR-Cas9 system, often compared to a highly precise pair of genetic scissors. This revolutionary method allows for the editing of DNA with a level of control that was once mere science fiction.

- **Guide RNA Design:**
- Crafting the guide RNA (gRNA) is like writing a GPS navigation route that leads Cas9 directly to the genetic address that needs editing. This gRNA must match the target DNA sequence perfectly to ensure accuracy and avoid detours that could lead to unintended genetic neighborhoods (off-target effects).

- **Target DNA Sequence Location:**
- The Cas9 enzyme, guided by the gRNA, scans the DNA, hunting for the exact sequence it's programmed to find. It's akin to using a metal detector that beeps only when it hovers over a very specific type of metal buried in the sand.

- **Molecular Mechanism of DNA Cutting:**
- When Cas9 and gRNA locate the target site, Cas9 performs a precise cut across the DNA strands, akin to a tailor making a careful incision in fabric. This cut effectively 'unzips' the DNA at that specific spot.

- **Cell's Repair Process Post-Cut:**
- The cell naturally attempts to repair this break using its internal toolkit. Two primary pathways, non-homologous end joining (NHEJ) and homology-directed repair (HDR), are like the cell's DIY repair guides. NHEJ can result in errors (mutations) at the repair site, while HDR uses a template with the correct sequence for precise repairs.

- **Gene Inactivation or Correction:**
- Depending on how the DNA is repaired, the gene can either be turned off (knocked out) or corrected to restore its normal function. It's similar to fixing a typo in an important email using autocorrect or by manually typing in the correct letters.

- **Accuracy and Avoiding Off-Target Effects:**
  - Ensuring the gRNA is impeccably matched to the target DNA is like proofreading an important text; one wrong letter can change the message entirely. Scientists are continually refining the CRISPR system to avoid these mistakes.

- **Strategies for Monitoring and Enhancing Specificity:**
  - Innovations in CRISPR technology include fidelity-enhancing mutations in Cas9 or truncated gRNAs, all to increase precision. Researchers are like app developers, updating the software to provide better service to users.

- **Real-world Applications and Clinical Trials:**
  - Currently, clinical trials are applying CRISPR to treat genetic disorders. For example, scientists are using CRISPR to target the defective gene in sickle cell disease, with the goal of restoring normal blood cell function. This is like sending in a specialized repair team to fix a recurring problem in a building's architecture.

Each aspect of CRISPR-Cas9 technology represents a step toward a future where genetic diseases may be treated with a few strategic snips and rewrites of our DNA. It's a powerful illustration of human ingenuity, turning a naturally occurring bacterial defense into a tool that may one day alleviate human suffering.

The ethics surrounding gene editing and stem cell use are indeed a complex terrain, a labyrinth where every turn and decision may have profound consequences. Like stewards entrusted with the delicate balance of a natural preserve, scientists and ethicists grapple with the responsibilities that come with such transformative capabilities. Stem cell research, with its immense promise, poses challenging questions about the manipulation of life's building blocks. Gene editing tools, especially CRISPR, allow for changes at the most fundamental genetic level, offering hope for cures but also raising issues of consent, access, and potential long-term effects on the human gene pool.

Much like conservation efforts that must weigh short-term gains against long-term ecological health, the scientific community must consider how the benefits of these technologies stack up against their ethical implications. In society, these tools raise the specter of 'designer babies,' the commodification of genetic traits, and new forms of inequality. Regulations lag behind the pace of scientific breakthroughs, creating a need for an ongoing and dynamic dialogue that considers not only the science but also the societal repercussions.

As we continue to unlock the power of genetics, it is imperative to tread thoughtfully, considering both the potential to heal and the obligation to preserve the integrity of human genetics. This conversation, while intricate and at times uncomfortable, is necessary for making informed decisions that serve the greater good and honor the trust placed in the hands of those who wield these powerful biological tools.

Let's take a deeper look at the ethical tapestry woven by the advancements in gene editing and stem cell research. Imagine these ethics as the rules of the road everyone must follow to ensure a safe journey for society.

- **Principles of Bioethics:**
    - Autonomy: Like having the personal freedom to choose your menu at a restaurant, autonomy insists on an individual's right to make informed decisions about their health, including gene editing.
    - Beneficence: This is similar to a lifeguard's duty at the pool; researchers must aim to ensure the well-being of participants, providing benefits while carefully navigating the risks.
    - Non-maleficence: Much like the principle 'first, do no harm' in healthcare, this rule ensures actions do not harm patients or others.
    - Justice: Imagine a line for a rollercoaster where cutting is not allowed; justice in gene editing means equitable distribution of the technology's benefits and its burdens without favoritism or discrimination.

- **Consent:**
    - This involves more than just signing a form. It's similar to agreeing to play a game only once the rules are clear. Patients must fully understand the potential risks and benefits of treatments arising from gene editing.

- **Access and Equity:**
    - The promise of these treatments should not be like a luxury car accessible only to a few. Everyone, regardless of socioeconomic status,

should have the opportunity to benefit from these medical advances.

- **Societal Implications:**
  - Genetic discrimination could be akin to being judged for a tattoo you were born with, undeserved and unjust. Alterations in human attributes might pressure individuals to conform to 'ideal' standards, much like unrealistic beauty expectations promoted in media.

- **Regulatory Frameworks:**
  - Current laws are like traffic lights guiding drivers; as the cars become more sophisticated, the traffic laws must adapt. Regulatory frameworks will need to evolve with these emerging technologies to ensure safety, ethical integrity, and public trust.

- **Public Engagement and Policies:**
  - Much as town halls gather community input on local decisions, public engagement is critical in shaping ethical gene editing policies. Interdisciplinary dialogue brings together diverse voices—patients, scientists, ethicists, policymakers—to address the complexities of these techniques and develop acceptable guidelines for their application.

Each of these elements is interdependent and crucial in navigating the ethical landscape of gene editing and stem cell research. They provide a foundation for informed, ethical decision-making that balances innovation with societal values. Through open dialogues and thoughtfully crafted policies, we can harness these powerful tools responsibly, ensuring that advances in biomedical science translate into fair and beneficial outcomes for all members of society.

Imagine genetics as a vast library that started with just a single, groundbreaking book—the pea plants studied by Gregor Mendel. Mendel's work laid down the ABCs of heredity, a foundation as crucial as the cornerstone of a magnificent building. Fast forward through decades filled with volumes of research, experiments, and discoveries, we arrive at the groundbreaking chapter penned by Shinya Yamanaka, who showed us how to turn back the pages of time. Yamanaka crafted a recipe to reverse mature cells, like those in our skin, into a state similar to the first blank page of life, known as induced pluripotent stem cells. This was like finding a way to erase the text of a novel and rewrite it in another genre entirely.

Between Mendel's foundational work with his simple peas and Yamanaka's sophisticated cellular alchemy, lies a vast compendium of stories, each contributing to the genetic narrative that informs our current scientific practice. As we turn each page, from DNA's double-helix discovery to the finer details of the human genome project, it's as if we've been filling in the atlas of human biology—one where every new discovery adds a layer to our map, guiding us to greater understanding and new horizons in medicine.

Here is the breakdown of the milestones in genetic research that have followed in the footsteps of Mendel's pioneering pea plant studies:

- **Identification of Chromosomes as Carriers of Genetic Material:**
  - It's as if we discovered not only that libraries have books, but that specific shelves hold the tomes that carry the stories of our lives. In the early 1900s, chromosomes were identified as the structures containing genetic information, crucial for passing traits from one generation to the next.

- **Discovery of the DNA Double Helix:**
  - In 1953, Watson and Crick unraveled the DNA double helix, the twisted ladder that scripts our biological narrative. This was akin to cracking the code that would allow us to read and interpret the vast volumes of genetic information.

- **Gene Cloning and Sequencing Technologies:**
  - The 1970s brought us gene cloning, giving us the ability to photocopy pages from the book of life, and DNA sequencing technology, which allowed us to meticulously transcribe those pages, letter by letter, into a format we could understand and analyze.

- **Polymerase Chain Reaction (PCR):**
  - PCR, developed in the 1980s, was like inventing a printing press for genes, rapidly creating copies of specific DNA segments for study and use, propelling research into a new age of genetic understanding.

- **The Human Genome Project:**
  - Initiated in 1990, this project was an international effort to map all the genes in the human genome. It's as if we composed a detailed atlas of our

genetic makeup, a tool that continues to guide research and therapy.

- **Rise of Gene Editing Tools:**
    - CRISPR and other gene editing technologies emerged as precise instruments, giving us the ability to alter the genetic text with intent and precision. It's like having a word processor for our DNA, equipped with the functionality to both backspace errors and insert corrections.

Each of these discoveries has contributed to our genetic compendium, forming a cascade of knowledge that has expanded our capabilities in genetic research and therapy. They have turned what was once pure observation into actionable insight, allowing us to venture into previously uncharted territories of medicine where we can tackle genetic diseases at their root. This ongoing genetic odyssey, with each advancement building upon the last, continues to shape our understanding of life itself, providing hope where it may not have previously existed.

Stem cell research stands on the cusp of dramatic advancements that could reshape medicine. As scientists grow more adept at understanding and manipulating these cells, the potential applications become both profound and transformative. Key areas include organ regeneration, where stem cells could be used to grow new, healthy organs in the body, eliminating the need for transplants from donors and reducing the risk of rejection. Additionally, the precise editing of genetic material holds promise for curing hereditary diseases by repairing the faulty genes before they can cause harm. Such breakthroughs would not only alleviate individual suffering but could also relieve healthcare systems by providing durable and potentially curative treatments. This burgeoning field is poised to revolutionize our approach to health and longevity, as mastery over the genetic blueprints of life unlocks therapies once imagined only in science fiction.

In conclusion, Chapter 3 has explored the pivotal role of genetic coding in the potential and versatility of stem cells. Understanding this intricate language is central to harnessing stem cells' full therapeutic potential, from repairing damaged tissues to treating genetic disorders. As we continue to decode the complexities of genetic information, we enable a future where personalized medicine and regenerative therapies become commonplace, fundamentally altering the trajectory of biology and medical science. This understanding is not merely an academic pursuit but a gateway to innovative treatments that could redefine healthcare's boundaries and enhance the quality of life for individuals worldwide.

CHAPTER 4 REGENERATION AND HEALING

You've now reached the exciting part of the world of stem cells—the body's own dedicated restoration specialists. Much like a well-equipped repair team, these remarkable cells have the extraordinary ability to replace damaged tissue and rebuild internal structures whenever and wherever necessary. Stem cells are vital to our body's maintenance and recovery, capable of developing into a myriad of cell types to restore functionality after injury or illness. Their pivotal role in health and disease offers a window into innovative treatments, promising to dramatically alter the landscape of current medical therapies. As you dive into this chapter, you will gain an understanding of how stem cells function as the architects and engineers of our body's continuous renovation, a dynamic process essential for sustaining life and improving wellbeing.

Stem cells are the foundational elements of the body's ability to regenerate and repair itself. They have two defining capabilities: self-renewal and differentiation. Self-renewal means that stem cells can divide and create copies of themselves over extended periods. This trait is like a master key that can create new keys indefinitely. Differentiation is the process by which these cells mature into specialized cell types with distinct functions, much like students deciding on their career paths and acquiring specific skills. Together, these abilities equip stem cells with the versatility to maintain and repair tissues throughout the body, from the blood cells that carry oxygen to the neurons that transmit information in the brain. This basic science of stem cells is at the heart of their promise in medicine, opening paths to potential cures and therapies that could dramatically change lives.

Stem cell self-renewal and differentiation is a complex, yet meticulously ordered biological process. Here's a straightforward guide to understand how it unfolds:

Step 1: Activation
- Activation is the first domino to fall in the chain reaction of stem cell activity. Stem cells remain inactive or quiescent until they receive the right signal, similar to an appliance that springs to life when plugged into an

electricity source. These initial triggers are often signaling molecules released by injured tissue or the stem cell's niche, which is its immediate environment.

Step 2: Cell Cycle Progression
- Once activated, stem cells enter the cell cycle, a series of phases where they prepare to duplicate their genetic material and eventually divide. Think of this like the preparation phase before an office printer starts churning out copies – it warms up, aligns the paper, and readies the ink.

Step 3: Self-Renewal
- Self-renewal is the process where a stem cell divides to produce at least one identical cell, ensuring the stem cell population is maintained. This is akin to a photocopier producing not just a copy of a document but also a replica of itself, thereby ensuring there's always a machine ready for future copying tasks.

Step 4: Differentiation Cues
- Differentiation cues are signals that prompt a stem cell to commence specialization. These can be certain proteins in the body that function like managerial instructions, telling an employee (the stem cell) to transition from a general role to a specific task within the corporation (the body).

Step 5: Specialization Pathways
- As a stem cell differentiates, it travels down one of many possible specialization pathways to become a specific cell type, just like a college student selecting a major to define their career path. This decision is directed by precise changes in gene expression – a series of molecular switches turning 'on' or 'off.'

Step 6: Final Transformation
- The final transformation is the completion of differentiation, where stem cells achieve their distinct identities – like a trainee emerging as a skilled professional. The cell has now acquired all the characteristics and functions necessary to play its part in the body.

Factors Influencing Self-Renewal and Differentiation:
- Stem cells are regulated by both internal mechanisms, such as their

genetic code that serves as an intrinsic blueprint, and external factors from their surroundings. Like a company adapting to market trends, stem cells respond to feedback from their environment to meet the body's needs.

Understanding this elaborate process provides us with the blueprints to potentially address a multitude of medical challenges. As we continue to elucidate each step in greater detail, the possibilities for intervention and therapy in conditions related to tissue damage and degeneration become increasingly tangible.

Embryonic stem cells are like the multilingual international diplomats of the cellular world. Just as a diplomat has the potential to interact with many different countries and cultures, embryonic stem cells can differentiate into virtually any cell type within the human body. They are derived from early-stage embryos and hold unlimited potential, making them incredibly versatile but also a subject of ethical debate.

Adult stem cells are more like local government officials; they have a more focused jurisdiction. Found in specific tissues like the bone marrow or the skin, they typically give rise to cell types within their original tissue. They're the specialists called upon for routine maintenance and local repair work, possessing a narrower range of abilities compared to their embryonic counterparts but are also less controversial in their use.

Induced pluripotent stem cells (iPSCs) are the retrained career-switchers. Scientists take mature cells, like those from your skin or blood, and reprogram them to act like embryonic stem cells. This process is akin to an experienced professional going back to university to retrain and embark on a completely new career path. iPSCs thus combine the limitless possibilities of embryonic stem cells with the practical and ethical advantages that come with using a patient's own cells.

Recognizing the uniqueness of each stem cell type helps us appreciate their potential roles in medicine, whether it's for broad-ranging therapies or targeted tissue repair. Each type has its function and fits, just like professionals in the workforce, contributing in different ways to the health and maintenance of our body.

Here is the breakdown of various stem cell types, their unique qualities, how we harness their potential, and the ethical factors we must consider:

- **Embryonic Stem Cells (ESCs):**
    - **Source**: Derived from the inner cell mass of a blastocyst, an early-stage embryo.
    - **Cultivation**: Grown in a lab dish, they require specific conditions to maintain their pluripotency.
    - **Genetic Traits**: Having the ability to become any cell type, akin to a master key for any lock.
    - **Regenerative Ability**: High potential due to their ability to differentiate into all cell types (pluripotent).
    - **Medical Use**: Studied for potential treatments for a wide range of diseases, including Parkinson's and diabetes.
    - **Ethical Considerations**: The use of human embryos raises significant moral questions and is subject to legal regulations.

- **Adult Stem Cells:**
    - **Source**: Located in tissues like the bone marrow, skin, and brain.
    - **Cultivation**: Isolated from tissue and grown in controlled lab conditions; more challenging to maintain.
    - **Genetic Traits**: Limited in the types of cells they can become, similar to having a key for only one lock.
    - **Regenerative Ability**: Primarily repair and maintain the tissue from which they originate.
    - **Medical Use**: Treatments such as bone marrow transplants for leukemia are well-established.
    - **Ethical Considerations**: Less contentious, as their use does not involve the destruction of embryos.

- **Induced Pluripotent Stem Cells (iPSCs):**
    - **Source**: Created in a lab by reprogramming adult cells, like skin cells, back to a pluripotent state.
    - **Cultivation**: Similar to ESCs but avoids the use of embryos, requiring specific factors to induce pluripotency.
    - **Genetic Traits**: Resemble ESCs in potential, like a blank slate that can be inscribed with any cellular role.
    - **Regenerative Ability**: Same as ESCs, with the theoretical potential to become any cell type.
    - **Medical Use**: Current research is exploring their use in regenerative medicine and drug testing.

- **Ethical Considerations**: iPSCs bypass many ethical issues associated with ESCs but research continues on safety and long-term implications.

By comparing the potential of each stem cell type to everyday roles and objects, we begin to see their place in both nature and medicine. They are not unlike resources in a community: some are versatile and can serve many roles broadly, while others serve in more specialized capacities. Understanding these distinctions not only fuels advancements in medical science but also shapes the ongoing ethical dialogue around how we utilize these powerful biological tools.

Stem cells play a crucial role in the body's repair process, operating through stages that mirror the body's natural healing response. When an injury occurs, stem cells in the affected area are activated, much like emergency responders at the scene of an incident. They begin dividing, in a process akin to a team of construction workers multiplying to tackle a large project.

As these cells multiply, some remain stem cells to continue the growth process—ensuring a steady supply of workers—while others start differentiating. Differentiation is the turning point where a stem cell commits to a specific function, similar to a construction worker deciding to be a plumber or electrician based on the job's needs.

These new specialized cells then replace the damaged tissue, integrating within the existing framework to restore the area much like renovating a damaged building to its original state. This regeneration can range from healing minor cuts to restoring complex tissues, such as liver cells regenerating after injury.

Throughout the process, the environment within the body provides signals to the stem cells, guiding their activity. These signals are comparable to a foreman at a job site directing workers, ensuring that appropriate tasks are completed at the right time and place for optimal repair.

As we learn more about this intricate process, we continue to uncover how the body orchestrates these stages of healing. New information may shed light on how to improve or accelerate regenerative therapies, offering hope for more effective treatments in conditions that currently have limited options for recovery.

Let's take a deeper look at how stem cells receive and act upon the repair calls of the body, a process that's as intricate as coordinating a space mission from mission control. Imagine each stem cell as a space vessel equipped with communication devices (receptors) ready to receive signals (growth factors and cytokines) from the control center (the stem cell niche).

1. Signal Reception: Like a satellite dish catching a specific television channel out of the countless signals in space, stem cell receptors bind to specific molecules designed to trigger them—these are typically proteins such as growth factors.

2. Signal Transduction Cascade: Once engaged, the receptors set off a relay race of molecular messengers inside the cell. This is much like mission control relaying critical information to the various departments within a space agency, each adding bits of information, refining the instructions before they reach the vessel.

3. Gene Transcription Changes: The final destination in this relay is the spaceship's 'command center'—the cell's nucleus. Here, the accumulated messages instigate changes in the crew's activity plan, or in cell terms, the transcription of specific genes. These instructions dictate which type of cell the stem cell will become.

4. Cellular Differentiation: Following the new blueprint, the stem cell embarks on its journey to specialization, much like an astronaut trains for a specific role in a space mission. It adapts, changes, and prepares its structure and function to fit its assigned task—whether that's becoming a blood cell, a neuron, or a skin cell.

5. Tissue Repair: As stem cells specialize, they begin to 'dock' into the damaged 'station' (tissue). Here, they integrate seamlessly with the existing cells, initiating repairs and over time, completely renovating the structure—turning a distressed site into a functioning part of the organism once again.

The environment—or the niche—a stem cell resides in, provides not just orders but the necessary support and conditions, akin to a training facility for astronauts, ensuring the stem cell has everything it needs for its mission success. Factors in this environment include supportive cells, extracellular matrix components, and a mix of signaling molecules, highlighting the collaborative nature of the body's repair processes.

Understanding this process is like mapping the journey from a signal initiation to the final repair of a spacecraft, revealing opportunities to enhance and support the repair mechanisms, which could mean faster and more efficient healing for the body.

Think of stem cells in therapy as the ultimate medical toolkit, packed with all the essentials needed for body repair. In treatments like bone marrow transplants, these cells are like a dedicated paramedic team, dispatched to the scene of the crisis—inside the patient's body. Bone marrow is the 'ambulance' that delivers these paramedic stem cells to the parts of the body where blood cells are made. Once there, they get to work, acting as the body's own repair mechanism to rebuild the blood cell supply. For patients with conditions like leukemia, this is similar to calling in a specialized disaster response team after a storm, helping to clear out the damage and restore essential services. This stem cell 'squad' meticulously replaces the destroyed cells and re-establishes a healthy blood supply, demonstrating just one of the life-saving ways stem cells are applied in medicine today.

Here is the breakdown on the intricate journey of stem cells during a bone marrow transplant:

- **Stem Cell Collection:**
  - Think of this as a precise fishing expedition, where only certain valuable fish are caught from the body's vast ocean. Stem cells are typically harvested from the bone marrow, peripheral blood, or umbilical cord blood.
  - Cells are isolated using a technique called apheresis, which can be likened to a sieve that catches only the needed cell types.

- **Preparation of Stem Cells:**
  - Much like selecting and prepping the best ingredients for a gourmet meal, stem cells undergo rigorous testing and conditioning.
  - They are often treated or 'coached' outside the body to enhance their performance once transplanted.

- **Transplantation:**
  - Picture this process as a carefully choreographed delivery service. Stem cells are infused into the patient's bloodstream, resembling a courier speeding to deliver a vital package to the correct address—the bone marrow.

- **Engraftment:**
  - Engraftment is the crucial 'settling in' period, where the stem cells home in on the bone marrow, similar to seeds rooting in fertile soil to grow.
  - Think of the stem cells now as construction foremen, assessing the damage and beginning the repairs on a compromised structure—the patient's immune system and blood cell production.
  - This phase can be fraught with challenges; the body's pre-existing cells may view the new cells as intruders, akin to neighborhood watch questioning new residents.

- **Restoration of Blood Cell Levels:**
  - Over time, the transplanted stem cells begin producing new blood cells, equipping the patient's body with fresh workers ready to construct and repair various tissues.
  - The production of red blood cells, white blood cells, and platelets is akin to a factory line resuming operation, working to churn out products needed for the body to function correctly.

- **Checkpoints and Mechanisms:**
  - Doctors closely monitor for signs of successful engraftment and complications, like a quality control team ensuring that the new products meet proper standards.
  - Immunosuppressants may be administered, akin to negotiators helping the incumbent citizens accept the new ones to prevent rejection or 'graft versus host disease.'

- **Recovery Process:**
  - The patient's recovery is a period of rebuilding. Much like a town recovering from a storm, support systems are in place to assist in the restoration of normalcy.
  - The patient undergoes tests and treatments which are comparable to regular inspections and maintenance work ensuring that the new infrastructure integrates well with the old.

This stepwise process highlights the synchronized dance of stem cells and medical intervention in renewing life at the cellular level. Each stage is a testament to the marvels of modern medicine, akin to a well-oiled machine that revives and maintains the body's internal ecosystem.

Stem cell research has been significantly advanced by the dedication of numerous pioneering scientists. One such individual is Dr. Shinya Yamanaka, who revolutionized the field by developing induced pluripotent stem cells, demonstrating that adult cells can be reprogrammed to an embryonic-like state. This breakthrough not only expanded our understanding of cell development but also opened new avenues for disease modeling and therapeutic possibilities, steering clear of ethical debates surrounding the use of embryonic stem cells. Personal accounts from patients who have undergone stem cell therapies further connect us to the real-world impact of this research. Their stories, from the lifesaving bone marrow transplants for leukemia sufferers to potential new treatments for previously incurable conditions, paint a vivid picture of the hopes and challenges inherent to this rapidly evolving field. These contributions and experiences underscore the transformative potential of stem cells not just in labs, but in lives across the globe.

The ethical debates surrounding stem cell research are akin to the rules and regulations that ensure urban development occurs responsibly and equitably. Central to the controversy are embryonic stem cells, which necessitate the destruction of a human embryo, prompting a moral examination of the beginning of human life and the integrity of research practices. As a result, guidelines have been established, much like planning permissions for city development, to navigate the moral complexities involved. These range from restrictions on the creation and usage of such cells, to consent protocols for donors, mirroring how public interest is protected in urban planning. On the other side of the spectrum, adult stem cells and induced pluripotent stem cells, while largely sidestepping these ethical hurdles, also have their use governed by stringent safety and ethical considerations to avoid unintended consequences. The conversation continues to evolve as breakthroughs in research, like the ability to create synthetic embryos, push us to regularly reevaluate and update these regulations, ensuring the field progresses in a manner that's ethically sound and socially acceptable.

Let's take a deeper look at the ethical tapestry woven by the advancements in gene editing and stem cell research. Imagine these ethics as the rules of the road everyone must follow to ensure a safe journey for society.

- **Embryonic Stem Cell Research:**
  - Embryos are often sourced from excess created during in vitro fertilization procedures, not unlike spare bricks that would otherwise remain unused at a construction site.
  - Consent is paramount, requiring the donors' informed permission, akin to homeowners agreeing to the repurposing of their leftover building materials.
  - Research scope is regimented globally, with some countries setting strict boundaries much like zoning laws that dictate what can be built and where.

- **Adult Stem Cell and iPSC Governance:**
  - Safety protocols for these cells are meticulous, similar to the careful maintenance checks done on a vehicle before a long trip to ensure the safety of passengers.
  - Ethical oversight includes ensuring donor privacy and monitoring cell manipulation, paralleling the careful regulations of personal data handling in the digital world.

- **Legislative Disparities:**
  - The legislative landscape varies as one might find with driving laws, which change from country to country, affecting how stem cell research can cross international borders.

- **Emerging Technologies and Ethical Adaptation:**
  - As research navigates new territories like synthetic embryogenesis, ethical frameworks are being stress-tested and recalibrated, much as engineers must update old infrastructure systems to meet new demands.
  - The philosophical debate continues to simmer, presenting questions as open and complex as discussions on the use of artificial intelligence in daily life.

Grasping these ethical elements allows us to appreciate the carefully thought deliberations behind each scientific venture, ensuring that the expansion of knowledge does not come at the cost of core human values. It's a matter of balancing the thrill of discovery with the gravity of responsibility, each advance measured and weighed to reflect the needs and morals of the global populace.

The horizon of stem cell therapy holds developments that might parallel the innovation seen in next-generation urban infrastructure and smart city design. Future stem cell treatments are poised to be more precise, personalized, and efficient, potentially offering cures to a range of diseases that are currently untreatable. This includes tailored organ regeneration, where stem cells could be used to grow replacement organs perfectly matched to the patient, thus eliminating organ donor shortages and the risk of rejection. Looking further, stem cell therapy could advance to a point where it is as integrated into health care as smart technology is in urban planning, optimizing the body's own mechanisms for self-repair and maintenance. This progress could transform our approach to medicine, representing a shift as significant as the transition from industrial cities to interconnected, technology-driven urban centers.

At the heart of this chapter lies the transformative power of stem cells, akin to the founding stones of a medical renaissance in regeneration and healing. Just as the invention of the printing press opened up a world of knowledge and progress, stem cells are paving the way for groundbreaking advances in medical treatments. They serve as the body's own repairmen, with the skill to rebuild and rejuvenate from within, offering new hope for diseases that were once deemed untreatable. Their role in tissue repair and organ regeneration signals the dawning of a new era where the ailments of the body can be addressed at their root, with precision akin to the intricate craftsmanship of a master watchmaker restoring a timepiece to its original glory. The implications of stem cell research stretch far beyond the laboratories and into the very fabric of human health, underscoring a future where the once unfathomable can become a reality. As we close this chapter, we not only understand stem cells more intimately but also the vast potential they hold—a potential that echoes the boundless aspirations of every visionary who ever dreamed of a better tomorrow.

CHAPTER 5 STEM CELLS IN RESEARCH

It's time to dive into Stem cell usage in research, bare with me as we have to re-cap just a little bit for context. Stem cells are the body's raw materials, the foundation from which all tissue and organs develop. They are unique in their ability to divide and become various cell types, a process known as differentiation. This capability makes them invaluable in medical research, offering potential treatments for diseases and conditions that currently elude cure. Stem cells hold the power to replace damaged tissues and may one day allow scientists to grow organs for transplants in a laboratory setting. As we venture through this chapter, you'll gain insight into how these vital cells can play a role in everything from regenerative medicine to personalized therapies, illustrating their far-reaching impact on healthcare.

Stem cells are the body's foundational units, akin to basic construction materials. They possess the remarkable capability to develop into an array of specialized cells that make up our organs and tissues, from the neurons in our brains to the muscle fibers in our hearts. This transformative ability underpins their role in medical research and the potential they hold for treating a host of conditions. Like master builders, stem cells can both replicate to maintain their population and adapt to form the specialized building blocks needed for repair and growth. Their versatility harbors the promise of new treatment avenues, offering hope for the regeneration of damaged tissues and the possibility of curing diseases that are currently beyond our reach. As we delve into their world, we uncover a landscape where medical science could be revolutionized, harnessing these microscopic powerhouses in the war against illness and injury.

Understanding how stem cells differentiate into specialized cells is a bit like following a recipe that can result in different dishes depending on the spices you add. Let's put on our chef hats and follow the process step by step.

First, a stem cell receives a 'wake-up call'—this is the initial activation that can come from various sources, much like a phone ringing to start your day. In the body, this signal could be a chemical from injured tissue telling the stem cell it's time to get to work.

Next, the stem cell answers the call and sets off a chain of events inside the cell, called signaling pathways. Imagine this as a master switchboard lighting up and sending messages to different parts of a building. Specific genes within the stem cell's DNA are turned on or off in response to these signals in a precisely controlled sequence—like a skilled chef knows exactly when to add certain ingredients to create the desired flavor profile.

As the genes turn on or off, the stem cell begins to change, taking on new characteristics and behaviors. The cell's structure might alter, like dough rising and taking shape in the oven, eventually leading to the final product—a specialized cell type such as a muscle cell or a nerve cell.

Critical checkpoints along the way ensure everything goes according to plan. These checkpoints are like quality control stations in a factory, designed to prevent errors and confirm that the cell is developing correctly at each stage of its journey. Regulatory factors, acting as supervisors, oversee the balance between keeping the stem cell as it is (self-renewal) and allowing it to become a new cell type (differentiation).

This balancing act is crucial for stem cell therapies aiming to regenerate damaged tissue or treat diseases. If we tip the balance too far towards differentiation, we might run out of raw stem cells. On the flip side, if we have too many stem cells without enough differentiation, they won't be able to repair the damage. It's akin to finding the perfect balance in a recipe between the base and the seasoning for the dish to come out right.

In stem cell-based therapies, such as regrowing a patient's skin after severe burns or restoring blood cells following chemotherapy, this control over stem cell differentiation is vital. It's like having just the right number of contractors to build a house—enough to do the job efficiently without overcrowding the site and causing confusion.

By maintaining a balance between self-renewal and differentiation, we harness the power of stem cells to create a range of treatments, potentially offering new life to tissues damaged by injury or disease. Just as a well-made dish can satisfy hunger and bring joy, correctly harnessed stem cells can heal and restore the body, underscoring the critical nature of understanding this complex, beautiful recipe of life.

Inside the body's biological toolbox, we find an assortment of stem cells, each with a special role to play. First, envision the embryonic stem cells as the Swiss Army knives, versatile and adaptable, capable of transforming into any cell type the body might need, from a brain cell to a heartbeat.

Next, we have the adult stem cells, the reliable screwdrivers in our kit. These are more specialized, typically fixing and maintaining the tissues they come from, like a heart stem cell that focuses solely on cardiac repairs.

Then there's the induced pluripotent stem cells, akin to a favorite power tool that's been recharged and repurposed. They start as regular adult cells, say, skin cells from your arm that scientists, through a bit of biological reprogramming, endow with the all-purpose utility of embryonic stem cells.

Each type of stem cell, with its distinct abilities and origins, plays a critical role in the growing field of regenerative medicine. Like a well-equipped toolbox, they collectively enable us to tackle a wide array of repairs within the body, from patching up damaged tissues to potentially reconstructing entire organs. This versatility is what makes the study and application of stem cells so promising and so crucial as we look towards healing the future.

Here is the detailed breakdown of the three main types of stem cells, each with its own 'specialized tool function':

- **<u>Embryonic Stem Cells (ESCs):</u>**
    - **<u>Characteristics</u>**: Can develop into over 200 different cell types in the human body. Prized for their pluripotency, they're the akin to the all-in-one multi-tool in our biological toolbox.
    - **<u>Applications</u>**: Potential to treat a myriad of diseases, form new organs for transplantation, and advance our understanding of human development.
    - **<u>Cultivation Methods</u>**: Harvested from the inner cell mass of a blastocyst, ESCs require very precise conditions to remain undifferentiated and grow.
    - **<u>Tissues Repaired or Regenerated</u>**: Could potentially be used to regenerate almost any damaged tissue in the body.
    - **<u>Directed Differentiation</u>**: Growth factors and other signaling molecules instruct ESCs on which types of cells to become.
    - **<u>Medical Procedures</u>**: Still largely in the experimental stage, they hold

promise for conditions like diabetes, spinal cord injury, and heart disease.

- **Adult Stem Cells:**
    - **Characteristics**: Compared to ESCs, adult stem cells are more like the trusty screwdriver—specialized and reliable for particular jobs. They are multipotent, meaning they can only become a limited number of cell types.
    - **Applications**: Used in bone marrow transplants and may also help treat diseases like leukemia and lymphoma, as well as conditions affecting the specific tissues they are sourced from.
    - **Cultivation Methods**: Found in small numbers in tissues such as bone marrow, brain, and blood vessels; isolated from the tissue and expanded in lab conditions.
    - **Tissues Repaired or Regenerated**: Predominantly geared to regenerate the tissues they originate from, like blood cells from bone marrow stem cells.
    - **Directed Differentiation**: The local environment or 'niche' provides cues to guide their development into specific cell types.
    - **Medical Procedures**: Routinely used in bone marrow transplants and also being explored for other disease treatments.

- **Induced Pluripotent Stem Cells (iPSCs):**
    - **Characteristics**: They are the comeback kings, akin to power tools that have been upgraded for a new purpose. Initially adult cells, they are reprogrammed back into a pluripotent state, giving them abilities similar to ESCs.
    - **Applications**: Like ESCs but without the ethical concerns, iPSCs could be used to model diseases, test drugs, and create tissues for transplantation.
    - **Cultivation Methods**: Scientists add certain genes to an adult cell, typically a skin cell, which reverts it back to a stem cell state.
    - **Tissues Repaired or Regenerated**: Technically, they can differentiate into any type of cell in the body.
    - **Directed Differentiation**: Treated with a specific cocktail of factors to guide them into becoming the desired cell type.
    - **Medical Procedures**: iPSCs are currently being used in clinical trials for conditions such as macular degeneration and Parkinson's disease.

Each stem cell type is a unique and essential element of our biological toolkit, offering a range of applications as diverse and promising as any high-tech gadgetry could. From repairing the body to modeling diseases and testing new drugs, understanding each 'tool' in depth reveals a future where medical treatment is not just about managing symptoms, but outright fixing the root problems.

Stem cells have a fundamental role in medical treatments by serving as replacements for damaged or diseased cells. Take bone marrow transplants, for example—a well-established therapy where stem cells are used to replenish the body's blood cell production. Here, healthy stem cells from a donor's bone marrow or blood are transplanted into a patient whose own blood-forming cells have been impaired, often due to chemotherapy or radiation. This process restores the patient's ability to produce blood cells and provides a functional immune system.

Looking ahead, these therapies hold enormous potential. Scientists are exploring how stem cells might be used to repair damaged organs or even grow new ones, a concept that's no longer relegated to science fiction. Imagine a patient with liver failure receiving a new liver grown from their own cells, eliminating the need for a donor organ and the risk of rejection. It's this potential for regeneration and disease treatment that makes stem cells a key focus for future medical breakthroughs, with the power to radically change how doctors approach the healing process. Each discovery and each successfully treated patient brings us closer to a future where the body's own building blocks can be harnessed to repair it from within.

Let's take a deeper look at the meticulous craft behind stem cell therapies, such as bone marrow transplants. Think of this process as akin to a well-coordinated ballet, where every movement is crucial and contributes to the enchanting final performance.

First off, harvesting stem cells is like gathering the ripest fruits from a tree. For bone marrow transplants, this involves a procedure where a needle is used to extract the marrow, which houses the stem cells, typically from the donor's pelvic bone.

Preparing these cells is then somewhat like a chef preparing ingredients, ensuring they're clean and suitable for the patient. This can include separating stem cells from the marrow and purifying them.

When it comes to matching donors with recipients, it's like finding a key that fits a lock perfectly, to minimize the chances of the body rejecting the new cells. Tests are performed to match blood types and human leukocyte antigens, which are proteins on cell surfaces that can trigger a response by the immune system if they're recognized as foreign.

The actual transplant reminds one of planting seeds in a garden. The prepared stem cells are infused into the patient's bloodstream, where they travel to the bone marrow and begin to grow, or engraft.

Mitigating rejection, or graft-versus-host disease, is like a diplomatic negotiation, where the body's natural defenses need to be convinced to accept the new cells. Medications that suppress the immune system are often the peacemakers in this scenario.

As for the frontier of organ regeneration, it's like the construction of a complex building. The stem cells must not only form the right types of cells but also create an intricate framework complete with blood vessels, known as vascularization, and integrate functionally with the body's systems.

Researchers have achieved breakthroughs, like creating mini-organs known as organoids, but challenges remain. One is ensuring that these new organs receive enough nutrients and oxygen through blood vessels, a process akin to laying down plumbing in a new construction.

Each step is part of a grand and noble effort to use the body's very own tools to rebuild and heal from within, and with every advancement, the complex jigsaw of regenerative medicine comes together a little more. This dance of discovery and innovation continuously pushes the boundaries of what we can achieve, bringing hope of new treatments that could one day mend hearts, heal spines, and save lives.

In the bustling metropolis of modern healthcare, stem cell research is akin to the avant-garde architects dreaming up the next generation of skyscrapers. Just as these thoughtfully designed structures transform the skyline, groundbreaking areas of stem cell study have the potential to redefine how we treat diseases and manage health.

For instance, researchers are exploring the use of induced pluripotent stem cells almost like developing buildings that can morph their structure according to need. These versatile cells, once ordinary body cells reprogrammed to their pristine state, could be the key to personalized medicine—tailoring treatments to the individual much as an architect customizes a building to its occupants.

Another pioneering field is the generation of organoids, miniature, simplified versions of organs. Picture a series of model homes, each representing a different style and layout. Organoids serve as test beds for drugs and disease study, allowing us to preview how an actual organ might respond to a treatment in the same way an architect's model demonstrates feasibility.

Then there's the quest to bioengineer whole organs for transplant. It's like constructing a brand-new sports stadium, with all the associated excitement and complexity. Here, scientists work not only to create the intricate cell framework of an organ but also to integrate it seamlessly with the body's own systems, laying down "plumbing" and "wiring" (blood vessels and nerves) that will allow the organ to come alive.

These architectural innovations within the realm of stem cell research are exciting but also laden with challenges, similar to how city planners navigate zoning laws and community needs. Nevertheless, as each puzzle piece clicks into place, we edge closer to a future where medicine is not only reactive but rebuilding, offering new hope and health where before there was none.

Here is the breakdown on the sophisticated process of shaping the future of regenerative medicine:

- **Creating Induced Pluripotent Stem Cells (iPSCs):**
  - **Step 1: Selection and Preparation**
  - Just as an architect selects prime materials for construction, scientists select mature cells, like skin cells, to be reverted to a pluripotent state.
  - **Step 2: Reprogramming**
  - Cells are then genetically 'renovated' using a series of reprogramming factors, much like installing new software to upgrade a computer.
  - **Step 3: Cultivation**
  - The new iPSCs are cultivated in controlled conditions—similar to

laying the foundation for a building—until they proliferate and form colonies.

- **Generating Organoids:**
  - **Step 1: Cell Aggregation**
    - Pluripotent cells, like iPSCs, are encouraged to form aggregates, comparable to gathering building supplies before starting construction.
  - **Step 2: Differentiation**
    - These aggregates are then exposed to specific conditions that induce them to start 'building' organoids, much like workers following blueprints to create different rooms or features of a building.
  - **Use in Drug Testing and Disease Modelling**
    - Like creating a miniaturized city model to test urban planning theories, organoids provide a scaled-down but accurate representation for scientists to test drugs and model diseases.

- **Bioengineering Organs for Transplantation:**
  - **Step 1: Scaffolding**
    - Just as an edifice requires a skeleton, bioengineered organs start with a scaffold that provides the structure upon which cells will grow.
  - **Step 2: Seeding**
    - Stem cells are then 'seeded' onto this scaffold, aligning precisely like construction materials on pre-marked plots of land.
  - **Step 3: Maturation and Integration**
    - With time and the right conditions, these cells mature into a functional organ, akin to a building reaching completion and becoming part of the city's landscape.
  - **Step 4: Vascularization**
    - Scientists also work on the vascular system—akin to the plumbing—so that the organ can receive the blood, oxygen, and nutrients it needs to survive in the body.
  - **Step 5: Transplantation**
    - The bioengineered organ is then seamlessly integrated into the patient's body, similar to how a modern addition to a traditional house must blend with the existing structure.

The strategies behind iPSC creation, organoid generation, and bioengineering are akin to building from the blueprints up, showcasing the beauty and intricacy of medical construction. These approaches are the pillars upon which groundbreaking therapies are built, standing as testaments to human innovation and our pursuit for healing in its most literal form—building back the parts that time or disease have worn down.

In the realm of stem cell research, ethical considerations are multifaceted and bear resemblance to debates on urban zoning—both involve decisions that reflect the values and needs of a community. Ethically, the use of embryonic stem cells raises significant concerns as it involves the destruction of an embryo, questioning the moral status of nascent human life. This is much like how urban zoning confronts dilemmas about the preservation of historical architecture versus the development of new structures.

Adult stem cells, though generally considered less contentious, still invite ethical scrutiny over donor consent and the equitable distribution of healthcare benefits akin to ensuring fair access to community resources when planning urban spaces. Furthermore, with advances like gene editing in stem cell research, concerns about long-term effects and potential for misuse mirror worries over the unanticipated impacts of building designs on city environments.

Another parallel can be drawn with the commercialization of stem cell therapies, mirroring the privatization of public spaces, which prompts questions about the commodification of health and the right to profit from biological resources. The evolving nature of societal norms is reflected in shifting attitudes towards these issues, influenced by emerging science and public discourse. As stem cell research progresses, it prompts ongoing evaluation of ethical frameworks, much like urban planning requires adaptability to changing societal needs and values.

Let's take a closer look at the intricate ethical landscape of stem cell research, where each decision reflects our collective values and principles much like zoning laws shape our communities.

With embryonic stem cell research, the crux of the ethical debate is the status of the embryo. Consider this: just as historic buildings are protected because of their potential cultural significance, embryos are valued by many for their potential to become fully developed humans, raising questions about when life begins and deserves protection under the law.

Informed consent for donors is no less important than getting a homeowner's permission before altering their property. For embryonic stem cells, this means ensuring that donors of embryos, often created for fertility treatment, fully understand their options and the implications of donating to science rather than, for example, another couple.

The legislative framework surrounding stem cell practices is as important as building codes in construction. Laws vary widely across the globe—some countries have stringent regulations akin to strict urban conservation rules, while others allow more freedom of exploration, creating a patchwork of practices similar to the varied zoning laws of cities.

Moving to adult stem cell research, issues of donor rights come to the fore, reflecting the principle of bodily autonomy as inviolable as property rights. Tissue ownership raises questions analogous to land titling—who owns the cells once they've left your body, and how are the profits from their use distributed?

As with any resource allocation, the distribution of medical treatments derived from stem cells must grapple with ensuring fairness and equity, much like city planners strive for even distribution of green spaces and public amenities.

Commercialization brings its complexities, mirroring the debates on public space privatization—how to balance the drive for innovation with ensuring access to treatments for all. The cost of new therapies can be as prohibitive as living in a gentrified neighborhood, and patenting biological materials and methods holds ramifications similar to the controversy over intellectual property rights in the software industry.

In sum, just as urban zoning encompasses various stakeholders' needs and values, ethical considerations in stem cell research demand nuanced deliberation and balance, aiming to respect the dignity of human life, the autonomy of individuals, and the equitable advancement of medical science.

Shinya Yamanaka's work in stem cell research stands as a monumental shift in the scientific landscape, akin to the historical moment when humanity first set foot on the moon. By discovering how to reprogram adult skin cells into induced pluripotent stem cells (iPSCs), Yamanaka paved a new path in medicine. His approach allows scientists to create stem cells without using embryos, circumventing the contentious ethical issues tied to their use. This breakthrough not only expanded our understanding of cellular development but also opened a gateway to innovative medical treatments, including organ regeneration and personalized medicine. Because of Yamanaka's contributions, the field of regenerative medicine now holds the potential for a future where many diseases could be treated more effectively, demonstrating how a single discovery can alter the trajectory of medical science.

As the chapter draws to a close, we recognize that stem cells bear the transformative potential to revolutionize the medical field, much like historical sea routes reshaped global trade and interaction. The ability of stem cells to differentiate, repair, and renew has profound implications for how we approach degenerative diseases, organ failure, and personalized medicine. They hold the promise of replenishing damaged tissues, offering hope for previously untreatable conditions, and driving advancements in medical therapies. Stem cell research constitutes not just the pursuit of scientific knowledge but also the quest for extending the quality of life and redefining the possibilities of health care delivery. The continued exploration of their capabilities could chart the course for a new era in medicine, echoing the monumental changes once brought about by the opening of new passages across the oceans.

CHAPTER 6 ETHICAL CONSIDERATIONS

Now in Chapter 6, 'Ethical Considerations,' where we navigate the complex moral landscape that accompanies the scientific advancement of stem cell research. Just as community leaders must carefully weigh the impact of their decisions on the well-being of their constituents, scientists and ethicists must consider how each advancement might affect not only patients but society as a whole. The potential of stem cell research to generate life-saving treatments must be balanced against concerns of bioethics, such as the status of embryonic cells, donor consent, and the equitable access to therapies. This chapter will explore the multifaceted ethical issues that stem cell research presents, examining how they guide the responsible progression of medical innovation and the governance of new technologies. Our goal is to provide clarity on the ethical framework that supports the field's integrity and future development.

The trajectory of ethical considerations in stem cell research has mirrored the rapid growth and significant breakthroughs within the field. In 1978, the birth of the first in vitro fertilization baby raised questions on the status of human embryos in science. By the late 1990s, with the isolation of human embryonic stem cells, this debate intensified, underscored by the potential of these cells to treat a myriad of diseases. In response to mounting ethical questions, the U.S. introduced the Dickey-Wicker Amendment in 1996, which prohibited federal funding for research that involved the creation or destruction of human embryos.

With advancements in the 2000s, culminating in Shinya Yamanaka's pioneering work on induced pluripotent stem cells, new regulatory guidelines emerged. The National Institutes of Health established clearer rules for federally funded research, balancing scientific discovery with ethical norms.

Internationally, the response varied with some countries embracing the research and others imposing strict limitations, reflecting diverse cultural and moral values toward embryonic research. Throughout the 2010s, as technologies like CRISPR emerged, ethical standards evolved to address concerns on gene editing. This includes guidelines by organizations like the International Society for Stem Cell Research, which continuously review ethical practices to adapt to the dynamic landscape of stem cell research.

Such milestones mark the maturation of stem cell research from a divisive topic to one bound by an intricate framework of ethical guidelines, reflective of society's commitment to harmonizing medical innovation with moral responsibility. This history encapsulates the collective efforts in ensuring that advancements provide health benefits while respecting the dignity and sanctity of human life.

The ethical journey in stem cell research began in earnest in 1978 with the birth of the first baby conceived through in vitro fertilization (IVF), which brought to the fore questions about the status of human embryos in science. For the first time, embryos that could lead to life existed outside the human body, prompting society to question how they should be used or protected.

This discussion reached a pivotal moment in the 1990s with the isolation of human embryonic stem cells. Research now promised medical miracles, treating previously incurable diseases. However, it also meant the destruction of these embryos, intensifying the ethical debate. The United States responded with the Dickey-Wicker Amendment in 1996, which prohibited federal funds for research that resulted in the destruction of human embryos, reflecting a societal attempt to draw a line between scientific progress and the protection of potential life.

The international community reacted diversely to such research—some countries, valuing the potential health benefits, permissively regulated embryo research, while others imposed tight restrictions, rooted in prevailing religious and ethical values. These varying approaches affected the global direction of stem cell research, influencing where scientists could conduct their work and how they could fund it.

In the 2000s, Shinya Yamanaka's research on induced pluripotent stem cells (iPSCs) offered an alternative, sidestepping many concerns by reprogramming adult cells to behave like embryonic ones. This innovation reshaped the ethical landscape, as it suggested a way to achieve similar research goals without using embryos.

The advent of gene-editing technologies, particularly CRISPR in the

2010s, propelled the ethical conversation forward. CRISPR's ability to precisely modify DNA raised possibilities of curing genetic diseases but also posed risks of unintended consequences and ethical breaches, such as germline editing.

In response to these advances, entities like the International Society for Stem Cell Research played a critical role, creating guidelines that respond to the latest scientific developments. These organizations work to strike a balance, incorporating diverse opinions and ensuring that stem cell research proceeds with broad societal support.

Legislation, technological progress, public sentiment, and ethical guidelines are intertwined in the narrative of stem cell research, each influencing the other. Together, they provide a mechanism for navigating the promise and perils of stem cell science, ensuring that as capabilities advance, ethical norms are not left behind. This balanced progression helps to maintain public trust in science, guiding the field responsibly into the future.

The ethical debates around embryonic stem cells can be likened to a city's discussions about whether to preserve a cherished historical building or make way for modern infrastructure. On one side, embryonic stem cells, derived from early-stage embryos, hold the potential to turn into any cell type, offering a real chance to repair damaged tissues and cure illnesses—picture this as a futuristic, state-of-the-art facility promising novel benefits to the city's residents.

Conversely, opponents see the embryo as the earliest form of human life, deserving of protection. In this analogy, the embryo resembles a historic building that holds value not for its utility, but for its intrinsic historical and cultural significance.

The crux of the debate, much like deciding whether to retain or replace the old with the new, centers on balancing potential future gains with respect for what we already recognize as valuable. These conversations weigh not just the immediate tangible outcomes—be it medical advancements or modern infrastructures—but also less quantifiable, ethical implications that resonate with our collective values. The choices made in each scenario reflect deeply held beliefs and ideals, demonstrating the careful consideration required when advancing into new but possibly contentious territory.

Here is the breakdown on the key arguments in the embryonic stem cell debate:

- **Support for Embryonic Stem Cell Research:**
  - **Medical Benefits:**
    - **Potential to Cure Diseases:** Just as architects design buildings to last and withstand diseases of structure, the adaptability of embryonic stem cells could lead to treatments for a range of chronic conditions, such as Alzheimer's, Parkinson's, and diabetes.
    - **Cellular Regeneration:** The same way city engineers might repair a decaying bridge, these cells have the capability to regenerate damaged tissues, potentially restoring function that has been lost to injury or disease.
  - **Scientific Advancements:**
    - **Understanding Human Development:** Embryonic stem cells give scientists a view into the earliest stages of human cellular processes, much like archaeologists uncovering the foundations of an ancient civilization.
    - **Research and Innovation:** With the potential of these cells, researchers can innovate on new medical treatments, similar to the development of new infrastructure facilitating modern city life.
  - **Impact on Legislation and Funding:**
    - **Enabling Progress:** Where legislation supports embryonic stem cell research, it often does so by framing these cells as a resource with the potential for transformative medical breakthroughs, akin to a city investing in cutting-edge public systems.
    - **Funding Scientific Exploration:** Just as city budgets allocate funds for public health and welfare projects, nations that support this research often channel resources into its complex exploration.

- **Opposition to Embryonic Stem Cell Research:**
  - **Ethical Concerns:**
    - **Respect for Potential Life:** For some, the early-stage embryo holds a status of potential human life, similar to how a heritage site holds symbolic significance for a city's history.
    - **Religious Beliefs:** Many religious groups view the embryo with the same sanctity as a person, much as a historical building might be seen as inviolable.
    - **Ethical Principles:** The principle of 'do no harm' is central to the opposition, resonating with the idea of not disrupting the existing moral fabric of a community.
  - **Impact on Legislation and Funding:**

- **Restrictive Laws:** Just as city ordinances might protect historic sites, laws in some countries protect embryos from research use by restricting or outright banning such activities.
- **Limited Funding:**
- **Public Discomfort:** Analogous to public outcry stopping the demolition of a historic landmark, public opinion has led some governments to limit or withdraw funding for embryonic stem cell research.
- **Philanthropic Priorities:** Much like benefactors choosing which civic projects to sponsor, private donors may direct funds away from embryonic research due to ethical concerns.

Understanding these facets of the embryonic stem cell debate helps us navigate the complex ethical territory, ensuring that our collective decisions in the medical domain are as informed and considered as those we make about our cities' skylines and historical legacies.

In stem cell research, the process of obtaining informed consent from donors is crucial, similar to the way property rights are observed when using someone's land. Just as you wouldn't build a house on someone's plot without their permission, researchers must first get consent from donors before using their biological materials. This ensures that the donors are fully aware of what the research entails, much like a landlord is made aware of the extent of construction before agreeing to it.

When a person donates their tissue or cells for research, they're entrusting a part of themselves to scientists with the expectation that it will be used with respect and for a purpose they agree with. It's similar to lending out personal items with specific instructions on their use; donors have the autonomy to set terms and are entitled to understand how their donation will contribute to scientific knowledge, resembling the way one understands how their property is to be used by another.

Informed consent isn't just a formality; it's a reflection of respect and autonomy in scientific practice. It acknowledges the donor's contribution to science while ensuring their agency and values are preserved, underscoring the need for transparency and dialogue in stem cell research. This practice underscores the ethical underpinning of the research community and deepens the trust within, much like observing property rights fosters goodwill in community relations.

Let's take a deeper look at the finely-tuned process of informed consent in stem cell research, unfolding it like a homeowner going through a meticulous home rental agreement.

- **Initial Donor Recruitment:**
- Imagine spreading word in the neighborhood that you're looking to rent out your space. Similarly, researchers reach out to potential donors through various channels, from clinics to public awareness campaigns, ensuring the donor pool is diverse and inclusive.

- **Communication of Information:**
- Before handing over the keys, every renter runs through the lease's terms with the landlord. In this step, researchers clearly explain the research's purpose, procedures involved, and any associated risks—just as a lease would outline what's expected of each party.

- **Comprehension:**
- Much like a landlord would need to confirm that the tenant understands every clause, researchers employ methods such as questionnaires or discussions to ascertain that donors truly grasp the implications of their contribution.

- **Voluntariness:**
- Ensuring donors are willing participants is like confirming a tenant isn't being pressured to sign the lease. Researchers take care to ensure participation is completely voluntary and free from coercion.

- **Documentation of Consent:**
- Signing a lease finalizes the agreement, just as donors formally document their consent, often in writing, confirming their informed and voluntary decision.

- **Ongoing Communication:**
- Throughout a tenancy, tenants may need clarifications or choose to renegotiate terms. Similarly, researchers maintain open dialogue with donors, who may have further questions as research progresses or may wish to withdraw their consent.

- **Special Populations:**
- Renting to someone from a different cultural background or to someone with limited housing experience requires extra care. Ethically, researchers take additional steps to ensure the informed consent process accommodates any cultural sensitivities and sufficiently supports individuals who might need more assistance to understand the research in which they're participating.

This comprehensive view of informed consent is akin to a well-crafted lease that respects the tenant's rights while ensuring they're fully informed. In research, it's a cornerstone that upholds the integrity of the scientific endeavor, honoring the donor's autonomy and cementing a foundation of mutual trust that's crucial for ongoing exploration in stem cell science.

When stem cell therapies hit the market, the question of ethical ramifications comes into play, much like the hot topic of privatizing public services arises within a community. Imagine a local park, once freely open to everyone, now charging an entrance fee after a private corporation takes over. Similarly, with the commercialization of stem cell therapies, treatments that might have been broadly researched in public institutions become intellectual property, potentially limiting access only to those who can afford them.

Such a shift could heighten disparities in healthcare, akin to how privatization of public utilities can lead to unequal services—imagine some neighborhoods having pristine tap water because they can pay premium rates, while others have to make do with less. The privatization of a public good like stem cell therapies could then lead to a tiered health system, where the wealthy enjoy access to life-enhancing treatments, while others are left behind.

Just as community members might argue that all citizens should benefit from public assets, there's a call within the medical community to ensure that life-saving therapies derived from stem cells remain accessible and equitable. The debate underscores the need to strike a balance between rewarding innovation and upholding the ethical commitment to healthcare as a shared value. It's a conversation not just about how these therapies work, but who they work for and the type of society we aspire to be—a society that champions innovation without compromising on inclusivity.

Let's take a deeper look at the marketplace of stem cell therapies, a world where scientific breakthroughs meet the economics of healthcare, almost like a bustling farmers' market where goods are as varied as they are valuable.

- **Drug Pricing Factors:**
- The cost to bring a stem cell therapy from lab bench to bedside is akin to a farmer's journey from seeding to selling. This price considers research and development expenses, clinical trial costs, and the often-complex manufacturing process, similar to a farmer's investments in high-quality soil and organic fertilizers.

- **Patent Laws:**
- Think of patents as recipe secrets for a farmer's prized homemade jam. They grant market exclusivity, which can prevent other companies from making generic versions and can drive prices up due to the lack of competition. Patents are the shields of the industry, protecting the intellectual property but also influencing accessibility.

- **Healthcare Policies and Insurance Coverage:**
- Healthcare policies and insurance can be imagined as community-shared programs that help make fresh produce from the market affordable for all. They decide which therapies are covered and the extent of co-pays, just like a community-supported agriculture (CSA) program might subsidize the cost of organic produce for its members.

- **Government Intervention:**
- Much like regulations ensuring fair trade at the market, government intervention in drug pricing can play a crucial role. It might set price caps or negotiate prices directly with pharmaceutical companies, striving to keep treatments within reach for citizens.

- **Potential Strategies for Accessibility:**
- **Tiered Pricing:** Imagining a sliding scale system wherein customers pay based on their ability to afford the goods, tiered pricing allows drugs to be sold at different prices in different markets or countries.
- **Subsidies:** Like coupons that help cover a portion of the costs for local shoppers, subsidies can reduce the financial burden on patients needing these life-saving treatments.

- **Public-Private Partnerships:** When a farmer teams up with local businesses to ensure everyone gets a taste of their harvest, it's a win-win. Similarly, collaborations between the private sector and government entities can spur innovation while ensuring public health interests are served.

How commercialization steers the priorities of scientific research can make the difference between a single-sourced market and a diverse one. Just as privatized markets can favor profits over variety, when therapies become commodities, the focus might shift from broad scientific inquiry to what's most commercially viable. It takes collective effort—governments, companies, and society—to blend the goals of advancement and accessibility, ensuring that the wealth of medical innovation fosters a healthier population rather than a deeper divide. It's about keeping the doors to the marketplace open for all, not just for those who can afford the entrance fee.

Legislation and regulatory bodies act as the rule-makers and watchguards for maintaining ethical standards in stem cell research. These entities define the 'rules of the game,' similar to how traffic laws govern the flow of vehicles on the road, ensuring safety and order. In the United States, for instance, the National Institutes of Health (NIH) sets guidelines that dictate what types of stem cell research can receive federal funding, much like how building codes determine what structures can be legally constructed.

These bodies also monitor compliance, require proper documentation, and consent protocols akin to how financial auditors review a company's accounts. The Food and Drug Administration (FDA) plays a pivotal role in reviewing and approving any new stem cell-based treatments before they can enter the market, ensuring they're safe and effective just as a health inspector would certify a restaurant's compliance with food safety regulations.

Their influence also extends to international collaboration—engaging with other countries' regulatory agencies to harmonize standards, much like how air traffic control towers coordinate to manage airspace globally. Through such efforts, they help to navigate the complexities of the research environment, address ethical dilemmas, and promote responsible scientific progress.

By setting clear standards and rigorously enforcing them, these legislative and regulatory bodies don't just preside over the current landscape; they actively shape the trajectory of stem cell research. Through a structured, scrupulous approach, they work to ensure that the rapid advances in this dynamic field are matched by thoughtfully developed and applied ethical principles.

The regulatory process for stem cell research is a meticulously structured journey designed to ensure safety, efficacy, and ethical adherence. Here's how it typically unfolds:

1. **Preclinical Research:**
    - Before any human trial, stem cell therapies are rigorously tested in the lab. This involves studying effects on cell cultures and animal models to predict how the therapy might behave in humans.
    - **Ethical Standards Assessed:**
        - Source of stem cells: ensuring they're ethically procured.
        - Animal welfare: adhering to guidelines for using animal models.

2. **Investigational New Drug Application (IND):**
    - Researchers file an IND with the FDA, presenting their findings and proposed human trials. This dossier must demonstrate the potential benefits outweigh risks.
    - **Documentation Required:**
        - Preclinical data.
        - Clinical trial design and protocols.
        - Manufacturing information ensuring the product can be produced consistently and safely.

3. **Clinical Trials:**
    - Conducted in phases, each step designed to answer specific questions about safety and effectiveness. From small initial groups (Phase 1) to larger populations (Phase 3).
    - **Ethical Standards Assessed:**
        - Informed consent: ensuring participants are fully aware of the risks.
        - Trial oversight: typically by an Institutional Review Board (IRB).

4. **Biologics License Application (BLA):**
    - Upon successful clinical trials, a BLA is submitted for the therapy to be approved for market. This is a comprehensive review of everything from trial data to labeling.
    - **Documentation Required:**
        - Detailed clinical trial results.

- Risk analysis and mitigation strategies.
- Plans for post-marketing monitoring.

5. **FDA Review and Approval:**
   - The FDA's team of experts scrutinizes the application to confirm the therapy's safety and efficacy.
   - **Ethical Standards Assessed:**
     - Data integrity: ensuring trial results are accurate and reliable.
     - Manufacturing standards: strict adherence to Good Manufacturing Practices (GMP).

6. **Post-Market Surveillance:**
   - Even after approval, therapies are monitored for any long-term effects or previously unnoticed issues. This phase involves gathering data from a wider patient base.
   - **Documentation Required:**
     - Adverse event reports.
     - Periodic safety updates.

7. **International Coordination:**
   - To bridge international regulatory discrepancies, agencies like the FDA collaborate with counterparts abroad.
   - **Reconciliation Strategies:**
     - Mutual recognition agreements; sharing trial data to prevent duplication.
     - Harmonizing guidelines through entities like the International Council for Harmonisation (ICH).

8. **Compliance Monitoring:**
   - Regulatory authorities conduct regular inspections and review of manufacturing practices, clinical trial conduct, and post-marketing data.
   - **Penalties for Violations:**
     - Can range from warning letters to fines or even revoking product approval.

This procedure ensures that stem cell therapies advance from conception to the clinic in a manner that's ethical, scientifically sound, and with patient safety as the paramount priority. It's a rigorous pathway that takes the discoveries of today and responsibly transforms them into the treatments of tomorrow, all the while upholding the moral imperatives that govern scientific inquiry and patient care.

As stem cell research strides forward, we're approaching a crossroads not unlike the one faced when GPS technology first enabled us to pinpoint our location down to the exact street corner. Just as GPS revolutionized navigation with profound implications, the potential of emerging stem cell technologies, such as gene editing, could redefine human health and biology. However, with these advancements comes a new set of ethical road maps we'll need to chart.

One potential future consideration is the concept of 'designer babies,' where gene editing might be used to select or enhance certain traits in offspring, akin to customizing features on a new car. This raises profound questions about fairness, as only some may be able to afford such enhancements, potentially widening the gap between socioeconomic classes. It also challenges the serendipitous nature of genetics, shifting it towards a more intentional, and potentially contentious, selection process.

Moreover, gene editing in stem cells could one day correct genetic disorders before a child is born, which is akin to repairing a bridge before a fracture can cause harm. While the benefits are immense, such intervention blurs the line between treatment and enhancement. It places us at the frontier of redefining what is considered 'normal' and what might be an 'improvement,' steering us into uncharted ethical waters.

We're just scratching the surface of the ethical implications as these technologies mature. While we cannot predict the future with certainty, it is crucial that we engage in thoughtful dialogue, much like a town meeting where community members voice their hopes and concerns. Only through open conversation can we navigate these upcoming ethical territories, ensuring that the future of human biology continues to unfold in a way that is both exciting and ethically sound.

Let's take a deeper look at the ethical crossroads ahead in stem cell research, diving into the fine details much like crafting an intricate hand-woven tapestry that reflects the complexities of our society.

- **Economic Disparity:**
  - Advancements in gene editing hold the potential to widen the socioeconomic divide. Imagine if only the affluent could afford the luxury of tailor-made genes, leading to a future where health and abilities could increasingly correlate with wealth, reminiscent of private schools versus public schools in education.

- **Medical Feasibility:**
  - The 'designer baby' concept, once science fiction, is veering closer to reality. As with any experimental technology, there's a gamut of medical risks and unknowns, comparable to testing a pilot jet with new technology. The decisions around deploying such technology weigh the benefits of disease prevention against the risks of unintended side effects.

- **Psychological Effects:**
  - The choice to edit genes may burden individuals with decisions that previous generations never faced, akin to navigating the complexities of an increasingly digital world where data privacy is a new concern.

- **Diversity and Genetic Norms:**
  - A focus on 'desirable' traits could shift the genetic mosaic of our populations, reducing the vibrant variety of humanity to a more homogenous palette. This might be similar to the way certain breeds of dogs are favored over others, potentially impacting biodiversity.

- **Role of Bioethics Councils and Impact Assessments:**
  - These organizations serve as humanity's ethical compass, assessing new technologies much like urban planners consider the environmental and social impacts of a new skyscraper. They provide guidelines meant to govern research practices and societal integration of new therapies.

- **Development of Ethical Standards:**
  - This would involve a sequence of steps: first by discerning the values we wish to uphold, akin to setting principals in law; then crafting guidelines, like writing the constitution for these technologies; and finally, implementing and enforcing them, which is similar to the way governments uphold laws.

- **International Conventions:**
 - Just as global treaties address worldwide concerns such as climate change, international conventions on gene editing seek to unify ethical standards across borders to prevent inequities and misuse.

- **Involving Public Opinion:**
 - In shaping the norms around stem cell technologies, the collective voice is paramount. Much as public referendums affect key policy decisions, public perspectives on gene editing must be considered via inclusive dialogue and education.

In conclusion, navigating the future of stem cell research ethics is an ongoing process that necessitates clear, transparent, and equitable frameworks to keep pace with technological advances. As we recall the tapestry analogy, it becomes apparent that each thread—each decision and regulation—is integral to the integrity and beauty of the final picture, which in this case, is a future society that is both scientifically advanced and ethically grounded.

In this chapter, we have tackled the ethical concerns that accompany stem cell research, which include the moral status of embryonic stem cells, the imperative of informed consent, the proprietary nature of commercialization, and the future implications of genetic manipulation. These themes underscore the responsibility of scientists, legislators, and the public to engage in ongoing, discerning dialogue as we navigate this evolving landscape. The intersection of ethics with stem cell research is not simply an academic cross-discipline—it is a reflection of our societal commitments and values as they pertain to the advancement of medicine and the respect for human dignity. The continuous evolution of stem cell technologies necessitates an equally dynamic ethical approach, ensuring that as our capabilities expand, they do so within the framework of thoughtful and conscientious discourse.

CHAPTER 7 STEM CELLS AND MODERN MEDICINE

It's now time we explore the groundbreaking realm of stem cell therapies—a medical revolution comparable to the advent of the internet. Stem cells, with their unique ability to develop into diverse cell types, are becoming the cornerstone of regenerative medicine, offering new hope for restoring damaged tissues and treating a myriad of diseases. Their potential to heal is not unlike how the internet opened up vast avenues for communication and information sharing, significantly altering our day-to-day lives. This chapter will navigate the function and importance of stem cell advances in modern medicine, delving into how they're reshaping treatment paradigms, much like the internet redefined global connectivity.

Stem cells are the body's raw materials, the cells from which all other cells with specialized functions are generated. In 1998, James Thomson, an American developmental biologist, was the first to isolate these cells from human embryos, presenting an unprecedented opportunity for medical treatments. The significance of this discovery is parallel to that of a fundamental scientific breakthrough, such as the splitting of the atom, due to the possibilities it opens up for the future. These possibilities include the regeneration of damaged tissues in the human body, offering new solutions for diseases and injuries that were previously deemed incurable or untreatable. Stem cells have the remarkable capability to divide indefinitely and become any of the body's cell types, providing a kind of cellular blank slate that can replace or repair damaged tissues, whether from spinal cord injuries, type 1 diabetes, Parkinson's disease, or heart disease. This ability is not merely a step forward in treatment options; it is a leap into possibilities of healing and recovery that were once considered beyond the reach of medicine. Each research advancement and clinical trial sheds more light on the practical applications of stem cells, potentially altering the treatment landscape much like how developments in computer science have reshaped modern technology.

The transition of stem cells from lab to patient is an intricate process paved with both promise and challenges. Here is how they journey from their versatile origin to becoming targeted therapies for diseases:

1. **Isolation**: First, stem cells are isolated; embryonic stem cells are derived from early-stage embryos, whereas adult stem cells are collected from tissues like bone marrow. Imagine the process as sourcing raw material, much like a craftsman would source wood for woodworking.

2. **Cultivation**: These cells are then cultivated in the lab, carefully nurtured under specific conditions that encourage them to multiply while maintaining their 'undifferentiated' state, similar to how a gardener fosters plant growth without shaping it.

3. **Differentiation**: Scientists then coax the cells to differentiate into the desired cell type. If, for instance, heart tissue is needed, the stem cells are exposed to conditions that mimic those of heart cells. This step is like training an apprentice in a specialized trade.

4. **Integration**: For the cells to treat an ailment, they must integrate into the body's existing tissue. In the heart example, these specialized cells would need to not only reside within heart tissue but also function in sync with it—analogous to integrating a new part into a machine without disrupting its operations.

5. **Clinical Trials**: Before stem cell therapies can be widely used, they must undergo rigorous clinical trials where their safety and effectiveness are put to the test. Researchers monitor patients closely, tracking not only if the therapy works but also any side effects that may arise.

6. **Ethical Oversight**: As research progresses, it must do so within an ethical framework. This is where the debate between the use of embryonic versus adult stem cells is most pronounced. Ethical bodies ensure considerations such as the welfare of donors, informed consent, and potential long-term effects are addressed.

7. **Approval**: If a stem cell therapy proves successful and ethical concerns are satisfactorily addressed, regulatory bodies like the FDA will approve it for use. Think of this step as getting the final stamp of approval from a quality inspector before a product hits the shelves.

8. **Post-Treatment Monitoring**: Even after the treatment is applied, patients are watched over an extended period to monitor the longevity and ongoing safety of the therapy, ensuring it continues to benefit without causing harm over time.

Throughout this journey, the ethical debate persists, especially in weighing the potential benefits against moral considerations. Each step in this process, from isolation to post-treatment monitoring, is guided by scientists' dedication to improving health, lawmakers' commitment to safeguarding ethical standards, and society's consensus on medical innovation's role in our lives. The potential for stem cells to treat diseases once deemed incurable is a testament to human ingenuity and an exemplar of modern medicine's evolving frontier.

In the garden of cellular biology, stem cells are like the green-thumb miracle workers, each with their special role in growth and healing. Embryonic stem cells are like seedlings; they're brimming with potential, ready to sprout into any type of plant you could imagine. This versatility means they can develop into any cell type in the body - but they require the right conditions to flourish.

Adult stem cells, on the other hand, are the steadfast, mature plants of the garden. They're settled into their identity - think of a rose bush or an oak tree - and they can replenish cells within their own type, such as blood or skin cells. While they're not as adaptable as the youthful seedlings, they're crucial for ongoing repair and maintenance.

Induced pluripotent stem cells (iPSCs) are the garden's ingenious grafts. Scientists take a mature cell and, with a bit of molecular magic, turn it back into a versatile seedling-like state. These iPSCs regain the ability to become any type of cell, similar to how a skilled gardener can coax a branch to grow roots and develop into a new plant entirely.

Each of these stem cells holds a different promise for healing. Embryonic and iPSCs offer a universal solution, a chance to replace any damaged tissue, while adult stem cells offer targeted, reliable repair. Understanding and harnessing their respective abilities is akin to mastering the art of gardening, where knowledge and care can turn a single seed into a flourishing garden, full of diversity and life.

Here is the breakdown of how embryonic, adult, and induced pluripotent stem cells transform from their raw form into medical miracles:

- **Embryonic Stem Cells:**
  - **Extraction:**
    - Extracted from the inner cell mass of a blastocyst, which is an early-stage embryo.
  - **Culture Environments:**
    - Grown in specialized conditions that mimic the natural embryonic environment, much like nurturing a tropical plant in a greenhouse.
  - **Differentiation Signals:**
    - Exposed to specific proteins and growth factors that direct them to become the desired cell type, akin to using a recipe to bake a specific type of cake.
  - **Application in Medicine:**
    - Injected into the body or grafted onto damaged tissues to promote healing, like seeding a barren field to grow a new crop.

- **Adult Stem Cells:**
  - **Gathering:**
    - Found in and harvested from tissues like bone marrow, fat, or skin; collected much like a farmer would harvest ripe fruit from a tree.
  - **Mobilization Techniques:**
    - May require certain drugs or procedures to be mobilized or increased in the bloodstream before collection.
  - **Differentiation Pathways:**
    - Less versatile than embryonic stem cells, they are typically coaxed into related cell types like blood or skin cells rather than any cell type.
  - **Therapeutic Uses:**
    - Replaced or injected back into the body to repair the tissue from which they originated, repairing damage like patching a hole in a favorite pair of jeans.

- **Induced Pluripotent Stem Cells (iPSCs):**
  - **Reprogramming:**
    - Adult cells are genetically reprogrammed to an embryonic stem cell-like state by introducing certain genes, similar to rebooting an old computer to its factory settings.
  - **Optimal Conditions for Growth:**

- Cultured in conditions that revert them to a pluripotent state, essentially rewinding their internal clocks.
  - **Directed Differentiation:**
    - After reprogramming, iPSCs are guided through the same differentiation process as embryonic stem cells, undergoing a makeover to become the cell type to treat diseased or injured tissues.
  - **Clinical Implementation:**
    - Being patient-specific, they hold immense promise for personalized medicine, like creating a custom key for a specific lock.

Each type of stem cell offers unique potential and faces its own set of challenges in the journey from lab to clinic. The process of coaxing these cells to become the treatments of the future is a complex dance of biology, medicine, and technology—with each step requiring precision, care, and a deep understanding of the blueprint of life. As we tap into this profound natural resource, stem cell research continues to promise groundbreaking therapies that may redefine our approach to healing and regeneration.

The ethical landscape of stem cell research is akin to a community debating the installation of public surveillance systems. Just as surveillance aims to enhance safety by monitoring for harmful activity, stem cell research holds the promise of significant medical breakthroughs, with the potential to combat devastating diseases and repair previously irreversible damage to the human body.

However, much like the cautious approach a neighborhood would take to ensure surveillance doesn't infringe on individual privacy, stem cell research must navigate the delicate balance between medical progress and ethical considerations. Questions concerning the source of embryonic stem cells, for instance, parallel discussions about the extent to which surveillance should be allowed to peer into personal lives.

On the other side, the use of adult and induced pluripotent stem cells offers a less contentious alternative, resembling community watch programs that rely on voluntary participation and transparency. The conversation is ongoing and multifaceted, involving scientists, ethicists, patients, and the public at large—all of whom must come together, much like a town hall meeting, to calibrate the balance between leveraging the life-saving potential of stem cells and upholding the values we hold dear in our society.

It's a dialogue where every perspective is critical, and every decision reverberates through the collective conscience, highlighting the need for both caution and enthusiasm as we chart the course of this remarkable field.

Let's take a deeper look at the intricate ethical tapestry woven by stem cell research, starting with the delicate threads of embryonic stem cell procurement. It's a process as finely balanced as the rights and responsibilities at play in public surveillance. When scientists harvest these cells, it's akin to navigating the blurred lines of individual privacy—in this case, the controversy centers on the origin of life and its intrinsic value, provoking a dilemma that tugs at the moral fabric of societies around the world.

Moving to adult stem cells, we find firmer ethical ground, similar to neighborhood watch schemes where participants opt-in, reflecting the voluntary nature of adult stem cell donation. Here, informed consent comes into play, comparable to signing a privacy notice that lays out how personal data will be used—a practice that's as essential in the lab as it is in cyberspace.

As we probe into the consent process, we find researchers beholden to a covenant of trust with both donors and patients, much like a custodian of personal information. They must navigate the complexities of fully informing donors—not only about the use of their cells but also about potential long-term impacts and rights, echoing the thorough transparency required in privacy agreements.

Oversight bodies like Institutional Review Boards (IRBs) and ethics committees wield the shears that shape the direction of research, ensuring each step respects individuals' rights and societal norms, with a vigilance parallel to that of a surveillance oversight committee safeguarding public interests.

Internationally, the legal frameworks for stem cell research are as diverse as the legislative landscape for surveillance technology—some nations embrace the technology with open arms, while others restrict it with an iron fist. Each culture, with its unique values and beliefs, threads its needle to stitch policies that may either enable or constrain the progress of regenerative medicine.

This field is in a constant state of evolution, with every new scientific triumph or ethical debate adding another layer to the complex quilt of medical innovation. It's a realm where personal ethics, societal values, and technological capabilities converge, demanding an agile adaptation as we collectively tailor the future of healthcare and human well-being.

Stem cells serve as a revolutionary tool in drug development, much like how a flight simulator serves aircraft engineers. Just as a simulator allows for the testing and refining of airplane designs under various conditions without leaving the ground, stem cells enable scientists to create models of diseases in the lab. These models mimic the characteristics of real human tissues and the mechanisms of diseases closely.

For example, to test a new medication for heart disease, scientists can induce stem cells to become heart cells that beat and respond like those in a patient's body. By applying the drug to these cells, researchers can observe potential benefits, side effects, and toxicity in a controlled environment before moving to actual clinical trials. This process provides a clearer and more accurate prediction of how a drug will perform in humans, optimizing the chances of success when the medication is eventually tested on people.

Using stem cells in this way expedites the drug development process and reduces reliance on animal testing, which often doesn't translate well to human responses. It's a shift towards more ethical, efficient, and precise drug discovery, paving the way for safer and more effective treatments. This innovative approach is not only transforming how medications are developed but also reflects the broader shift towards personalized medicine, where treatments can be tailored to the individual characteristics of a patient's disease.

Let's take a deeper look at the process of coaxing stem cells to take on the role of heart cells in the symphony of drug testing. Imagine you're a director of a play, and your job is to guide actors—here, the stem cells—into embodying their roles as convincingly as heart cells.

Firstly, the culture of stem cells is like casting. Cells are carefully selected and placed in an environment—a bioreactor that's akin to a rehearsal space—where they are nurtured with the right nutrients and growth factors. These factors act like the director's instructions, guiding cells on how to divide and develop. For cardiac differentiation, we introduce specific agents such as Wnt proteins and growth factors like BMP4, which serve as cues for the cells to start specializing, much like an actor learns to adopt mannerisms of their character.

Once the stem cells receive these cues, they embark on a transformation, adopting the identity of cardiac cells. This change is meticulously monitored, akin to watching daily rehearsals to ensure the actors' performances align with the heart cells' behaviors—beating in unison, responding to stimuli, and exhibiting electrical activities specific to heart tissue.

In this theater of drug testing, we now introduce our experimental compounds—the potential medications—to see how they affect these newly minted cardiac cells. Just like reviewing a dress rehearsal, scientists measure cell viability, electrical activity, and contraction rates to evaluate the drug's impact. It's a live screening where the drugs' effects on these heart cells are tested for potential benefits and adverse reactions, offering a real-life preview of clinical performance.

Compared to traditional animal testing, this stem cell model stands out like a high-definition simulation, providing a human-specific context that can often predict human responses more accurately than animal models.

On the regulatory stage, there's a script of guidelines that must be followed, ensuring ethical provenance of stem cells, informed consent for donated materials, and compliance with safety regulations. It's the moral compass that aligns scientific progress with ethical standards, ensuring the audience—the patients and medical community—can trust the final act.

Finally, this method plays a leading role in the narrative of personalized medicine, tailoring the treatment to the patient's unique genetic script. By using the patient's own cells, it's akin to customizing the play to the individual down to the finest detail, ensuring the performance addresses their specific medical needs with extraordinary precision. This is the promise of stem cell research in drug development—a promise that turns the page towards more effective, personalized healthcare.

Stem cells are pushing the boundaries of medical treatment into the realm of personalization, much like how modern video games allow players to tailor their gaming experience. Just as a gamer can create a character with distinct abilities, appearance, and even narrative choices to suit their play style, doctors could use stem cells to develop treatments that are tailored to the patient's unique biological characteristics.

Imagine you are customizing your character in a role-playing game. You select skills and equipment that will help you in the quest ahead. Likewise, stem cells can be guided to become any specific type of cell that a patient may need, whether that's insulin-producing cells for someone with diabetes or cardiac cells for someone with heart disease — the customization is patient-specific.

In this transformative gaming experience, each choice affects the progression through the storyline; in medicine, each stem cell-derived therapy reflects the patient's genetic makeup and medical history, enhancing treatment effectiveness and reducing the risk of adverse reactions. The end goal is much the same: an enriched, tailored experience — in gaming, it means a more immersive play, and in medicine, a healthier life.

This is the vision of personalized medicine with stem cells at its core, providing a bespoke approach to healing — a future where treatment is as customized as a player's avatar, designed to fit the narrative of each individual's health journey perfectly.

Here is the breakdown of the meticulous craft that is personalizing medicine with stem cells, laid out as clearly as a blueprint for building your dream home:

- **Genetic Mapping:**
  - Much like an architect surveys the land before drawing up house plans, doctors begin by mapping the patient's genome. This gives a detailed layout of genetic sequences, identifying any variations that could impact treatment.
    - **Sublist:**
      - Blood or tissue samples are collected from the patient.
      - High-throughput sequencing is then used to catalog the genetic data.
      - Bioinformatics tools analyze this data, highlighting genetic markers relevant to the patient's condition.

- **Stem Cell Differentiation:**
  - After the genetic groundwork is laid, stem cells act like raw materials ready to be shaped. Depending on the genetic instructions, they are coaxed—like clay in a potter's hands—into the cell types the patient needs.
  - **Sublist:**
    - Stem cells are cultured in a lab environment with the appropriate growth factors.
    - Researchers monitor the cells, guiding them through stages of development mirroring natural growth.
    - Environmental cues are adjusted, such as oxygen levels or pressure, to enhance differentiation.

- **Tailoring Treatment:**
  - Once the cells have fully taken shape, they're crafted into a treatment as bespoke as a custom-tailored suit. Each therapy is fine-tuned to snugly fit the patient's unique genetic makeup.
  - **Sublist:**
    - Cells are often engineered to possess or lack specific features according to genetic markers.
    - Laboratory testing ensures the cells behave expectedly and beneficially.
    - Dosages and administration methods are personalized.

- **Role of Bioinformatics:**
  - The work of bioinformatics in this process is like the behind-the-scenes tech wizardry that makes a smart home efficient and responsive. It sifts through genetic data to find the best plan for stem cell application.
  - **Sublist:**
    - Bioinformaticians use computational algorithms to draw connections between genetic traits and stem cell properties.
    - Predictive modeling forecasts how stem cells will respond when manipulated.
    - Databases of genetic information help identify possible adverse reactions to therapies.

- **Increased Efficacy & Reduced Side Effects:**
  - This level of customization ensures the treatment hits the bullseye, increasing the chance of success and decreasing the potential for side effects—much like a well-organized daily planner fine-tunes your schedule to

avoid conflicts and maximize productivity.
- **Sublist:**
- Customized stem cell therapies lower the risk of immune rejection.
- They increase the precision of attacking diseased cells without affecting healthy ones.
- The detailed genetic analysis informs ongoing adjustments to treatments as the patient's condition and needs evolve.

In summary, personalizing medicine with stem cells is a process of turning the raw material of human potential into treatments as unique as the patient's DNA. It's a harmonious blending of biology, technology, and artistry, all working together to script a healthier future, scene by scene, cell by cell.

As we look toward the horizon of biomedical advancement, stem cell technology stands out as a field poised for revolutionary change, not unlike how jet engines transformed airplanes and reshaped the very concept of global travel. In the near future, stem cells could dramatically alter the landscape of organ transplantation. Today, finding a compatible organ donor is a race against time, with many hurdles including limited organ availability and the risk of rejection.

Stem cells, however, could unlock the ability to grow organs in the lab – imagine a patient's own cells used to create a personalized heart or kidney, reducing the risk of rejection to almost zero. This isn't just repairing damaged tissues; it's about building them anew. In this envisioned future, waiting lists for transplants could be much shorter, and the need for lifelong immunosuppressants following surgery could be a thing of the past.

Researchers are already laying the groundwork for such breakthroughs, meticulously orchestrating the environment and stimuli that coax stem cells to form complex structures of specific organs. The process is delicate and precise, requiring a thorough understanding of the cell's development.

Just as the aerospace industry overcame barriers in physics and engineering to allow flights to become faster, safer, and more efficient, stem cell research is making strides toward surmounting biological constraints, aiming to shift organ transplantation from a system of scarcity to one of abundance. This is the potential future of stem cell technology – one where the skies of medical possibility are wide open.

Let's take a deeper look at the intricate process of growing organs in the lab using stem cell technology, a process that's as complex and meticulous as the art of watchmaking. We start with stem cells, the all-purpose timekeepers of the body, which like the base components of a watch, can be shaped into any part necessary.

- **Isolation and Preparation:**
   - The first step is like selecting the finest materials for a custom timepiece. We start by isolating a patient's stem cells—usually from bone marrow or blood.
   - These cells are then meticulously cleaned and prepared, establishing an uncontaminated and viable population for the next stage of development.

- **Cell Specialization:**
   - Much like watch parts are specialized for specific functions, stem cells are coaxed into becoming particular cell types—an intricate process known as differentiation.
   - Growth factors and various signaling molecules act as the specialized tools to precisely shape these cells, guiding them down the path to becoming tissue-specific, such as hepatocytes for liver or cardiomyocytes for heart.

- **Scaffolding Techniques:**
   - To grow a three-dimensional organ, cells need a frame, much like the framework that supports a building during construction. This is provided by biocompatible scaffolds, which are designed to mimic the extracellular matrix—a structure that supports cells in an organ.
   - These scaffolds guide the growth and organization of the cells into the complex architecture required for a functioning organ.

- **Bioreactor Conditions:**
   - Just as a seed needs the right conditions to grow, stem cells require an environment that closely simulates the body's natural conditions. Bioreactors offer this, providing controlled temperature, nutrients, and gases to the developing organ.
   - The conditions within the bioreactor are fine-tuned to encourage cells to mature, connect, and form functioning tissue.

- **Ensuring Functionality:**
  - Before a lab-grown organ is ready for transplantation, it must be rigorously tested—similar to a watchmaker ensuring that all parts of the timepiece are working harmoniously.
  - This involves assessing functions such as electrical activity in heart tissue or filtration in kidneys to ensure that the organ will perform its intended role in the body.

- **Addressing Challenges:**
  - One of the main hurdles is ensuring the organ receives enough nutrients and oxygen, akin to the challenge of evenly winding a watch to keep it ticking. This is addressed through the process of vascularization, creating blood vessel networks within the organ.
  - Organ maturation is like the fine-tuning of a timepiece, ensuring it keeps time accurately over long periods. Researchers work to ensure the organ's long-term viability and functionality.
  - Post-transplant integration is crucial. It can be compared to fitting a new gear into an existing watch; the organ must not only fit but work in synchrony with the patient's body.

This whole process is groundbreaking, turning the potential of stem cells into tangible hope for many who wait for transplants. The promise is a future where organ shortages are no longer a constraint, and the treatment is personalized to every patient's needs—heralding a new era in medical science as transformational as the flight in aerospace, leading us to a time of tailored remedies and renewed lives.

Stem cells are a cornerstone of regenerative medicine, similar to how digital technology has become a foundation of modern life. They hold the promise of repairing damaged tissues, modeling diseases for better drug testing, and potentially offering cures where none existed before. Their ability to develop into various cell types makes them invaluable for understanding and treating a range of health issues, from genetic disorders to the aftermath of injuries. As research advances, the role of stem cells is only expected to grow, opening doors to innovative treatments and enhancing the quality of life for patients worldwide, continuing to weave their way into the fabric of healthcare much like digital advancements have woven into daily routines.

CHAPTER 8 FUTURE HORIZONS

Venturing into the realm of stem cells and their future in medicine is akin to peering through a telescope at the expanding universe. Stem cells, with their remarkable ability to transform into a diverse array of cell types, represent a frontier in medical science that is vast with potential. Currently, they play a critical role in regenerative treatments, repairing damaged tissues, and providing hope for incurable diseases. Moving forward, the advances within this field may revolutionize not just individual treatments but the entire healthcare system, potentially leading to personalized therapies, reduced transplant rejections, and innovations yet to be imagined. Just as space exploration challenges our reach to the stars, stem cell research challenges the limits of human health and longevity.

Stem cells are unique units in the body, akin to blank slates, with two defining capabilities: self-renewal and differentiation. Self-renewal allows them to divide and replicate for an extended period while maintaining their original state. Differentiation is the process by which these cells transform into specific cell types with specialized functions. Think of stem cells as the body's multipurpose tools, capable of regenerating and replacing damaged tissues, an ability that's key to the body's repair system and holds immense potential in medical treatments. Their versatility is a central focus of current research, offering insights on how to tackle diseases that today remain without a cure, striving towards the ultimate goal of healing at a cellular level.

Stem cells come in various forms, each with specific sources and potential applications in the medical field. Among these are embryonic stem cells, adult stem cells, and induced pluripotent stem cells (iPSCs).

Embryonic stem cells (ESCs) are derived from early human embryos. Their incredible flexibility—being able to become any cell type—makes them a powerful tool in research. For instance, ESCs have been used in clinical trials looking to restore sight to those with retinal diseases.

Adult stem cells, found in tissues like bone marrow and skin, are more specialized. They can't turn into any cell type but are critical in everyday healing processes. Bone marrow transplants, a common treatment for leukemia, are possible due to the self-renewal and differentiation abilities of hematopoietic (blood-forming) stem cells found in the marrow.

Induced pluripotent stem cells (iPSCs) are a scientific marvel. Ordinary adult cells are reprogrammed back to an embryonic-like state, regaining the versatility to develop into various cell types. This has huge implications for personalized medicine—using a patient's own cells to treat or study their disease.

The process of stem cell self-renewal and differentiation goes somewhat like this:

1. A stem cell divides, producing two cells.
2. One cell remains a stem cell, able to divide again; the other begins to differentiate.
3. The differentiating cell receives signals from its environment—chemicals and proteins that guide its development.
4. Over time and through various stages, it specializes, acquiring specific functions, such as conducting electricity in the heart if it becomes a cardiomyocyte, or processing information in the brain as a neuron.

Recent advancements in stem cell research have made headlines. From bioprinting tissues with stem cells to using iPSCs for drug testing and disease modeling, the scope is broadening. Challenges remain, including understanding the mechanisms behind differentiation and avoiding the risk of cells turning cancerous. Yet, the sheer potential of stem cells gives hope for future therapies that could regenerate damaged organs, treat a myriad of diseases, and fundamentally change healthcare.

Imagine stem cell therapies as the emergency repair kits of the medical world. Just as you might use a repair patch to fix a leak in a bike tire, doctors can use stem cells to repair damaged tissues in the body. For example, in the case of a bone marrow transplant, stem cells from a healthy donor are like a fresh set of batteries inserted to re-energize a failing flashlight—the body's immune system. This can be life-saving for patients with leukemia, where their own blood-forming cells have malfunctioned. These therapies are not without their challenges, but just as the right patch can make a tire good as new, the right stem cell treatment has the potential to restore a patient's health, signifying the tremendous promise that these therapies hold.

Here is the detailed breakdown on the journey of stem cell transplantation, explained as simply as a friend walking you through a step-by-step recipe for a gourmet dish:

- **Donor Matching:**
  - Like finding the perfect seasoning that complements a meal, doctors must find a stem cell donor whose tissue type matches the recipient's as closely as possible. This is pivotal to reduce the risk of rejection.
    - Genetic screening tests, such as HLA typing, are performed.
    - A database search identifies potential donor matches.
    - Further compatibility testing is done to ensure the best match.

- **Stem Cell Harvesting:**
  - Extracting stem cells is akin to gathering the freshest ingredients. The type of stem cell required—whether from the bone marrow, peripheral blood, or umbilical cord blood—determines the method of collection.
    - Bone marrow is collected under anesthesia through a needle puncture.
    - Peripheral blood stem cells are mobilized into the bloodstream and collected via apheresis.
    - Umbilical cord blood is gathered immediately after birth with no harm to mother or child.

- **Recipient Preparation:**
  - Just as one might prepare a kitchen for cooking, the recipient's body must be prepared to receive the new cells. This involves creating space within the bone marrow and preventing immune rejection.
    - Patients undergo chemotherapy and/or radiation to eliminate diseased cells.
    - Medications may be used to suppress the immune response.

- **Stem Cell Transplantation:**
  - Introducing the stem cells into the patient's body can be likened to adding a key ingredient to a dish that brings it all together.
    - Stem cells are infused through a central venous catheter, similar to a blood transfusion.
    - Cells travel to the bone marrow, beginning the integration process.

- **Post-Transplant Recovery:**
    - After the transplant, the monitoring period is crucial, much like letting a dish rest to fully develop its flavors.
    - Patients are closely observed for signs of engraftment, where new blood cells start to form.
    - The immune system's response is monitored for complications, including infection or graft rejection.

- **Dealing with Graft-Versus-Host Disease (GVHD):**
    - GVHD is when donor cells attack the recipient's tissues; akin to a spice that overpowers a dish, this must be managed carefully.
    - Preventive medications may be used to keep GVHD in check.
    - In case GVHD occurs, treatment options include steroids or other immunosuppressants.

This process is intricate, with each step tailored to the individual to enhance the chance of successful treatment while minimizing risks. The art of stem cell transplantation holds a wealth of potential, similar to a complex recipe that, when followed with care and precision, can result in a life-saving meal for those in need.

Future innovations in stem cell medicine are bridging the gap between the fantastic stories of science fiction and our everyday reality. Picture a world where, much like mechanics repair and replace worn parts of a car, doctors could use stem cells to replace damaged tissues in the body—a concept once relegated to the pages of a sci-fi novel. These aren't just any replacement parts; they're customized to the individual, reducing the risks of rejection and side effects. Think of a patient with heart disease receiving new, healthy heart cells grown from their own tissue, or someone with a spinal cord injury regaining mobility thanks to stem cells that repair the damaged neural connections. This is where the narrative of science fiction meets the tangible progress of science fact, presenting a future where healing is not just about treating symptoms but restoring health at the cellular level, offering new chapters of hope for patients who once faced untreatable conditions.

Let's take a deeper look at how the seed-like pluripotent stem cells are nurtured to sprout into a variety of tissue types in the fertile ground of the lab. The process is fascinatingly similar to that of a masterful gardener who not only cultivates plants from seeds but also guides them into growing into specific varieties.

- **Extraction of Stem Cells:**
    - The journey begins with the stem cell harvest, akin to collecting the seeds of potential from the biological orchard that is an embryo or a tissue sample.
    - These stem cells are then isolated, much like picking the choicest seeds for the best outcomes.

- **In-Lab Cultivation:**
    - Next, the stem cells are planted in the nutrient-rich soil of a culture medium, providing them the exact nutrients needed to grow and thrive.
    - Just as a gardener monitors environmental conditions, researchers carefully control temperature and other factors that might influence the stem cells' growth.

- **Differentiation Process:**
    - With care and precision, akin to training a vine to climb a particular trellis, scientists guide the stem cells to turn into specific tissue types through meticulously timed biological signals.
    - This is arguably the most elaborate part, where they add substances that act like sunlight and rain, coaxing these universal cells down different paths—be it into heart muscle cells or neurons.

- **Transplantation:**
    - Once matured, these cells are ready for transplantation—the transplant analogous to repotting a mature plant into a garden where it's needed.
    - Like checking the forecast before a plant's transition outdoors, compatibility tests are run to ensure the patient's body is ready to accept these new cells without rejection.

- **Ongoing Challenges:**
    - Ensuring the precision of cell differentiation is akin to a gardener making sure a fruit tree produces the desired fruit; it requires exactness and consistency.
    - Preventing immune reactions is like protecting a transplant from pests; both involve safeguarding the integrity of the new addition.
    - Scaling up lab-grown tissues is similar to a small-scale farmer expanding their operations; it requires innovation and efficient processes to meet larger demands without sacrificing quality.

With the ethos of a master gardener and the precision of a scientist, researchers are turning the science fiction vision of custom-grown biological repair kits into reality, forging a future where healing comes naturally and treatment is truly tailored.

The ethical dimensions of stem cell research are as intricate and vital as in any advanced scientific field, where innovation meets profound moral questions. In fields like artificial intelligence, for instance, we grapple with the ethics of autonomy and privacy. In a similar vein, stem cell research demands a balance between the promise of medical breakthroughs and ethical respect for the origins of these cells, particularly when it comes to embryonic stem cells. Decisions about how and when to use these cells necessitate a dialogue that considers the potential for life-saving treatments against concerns for the respect for embryonic life. Moreover, issues of consent for donated cells and the fair distribution of therapies derived from stem cells reflect the complexities seen with organ donation and access to healthcare technology. Such considerations require clear guidelines and thoughtful debate to navigate the delicate terrain between scientific progress and ethical responsibility. As we deepen our understanding of stem cells, the importance of these ethical discussions grows, ensuring that the path forward is both scientifically sound and morally grounded.

Let's take a deeper look at the ethical tapestry woven into every fiber of stem cell research, unraveling its complexities over a friendly chat. Think of embryonic stem cells as the seeds of a rare plant that holds the cure for countless ailments. Here lies the crux of ethical debate: to harvest the seeds, the plant must be destroyed. Ethical decision-making here juxtaposes the moral status attributed to embryos with the transformative health benefits stem cell research promises. It's akin to deciding between preserving an ancient tree and curing a village's persistent illness.

Informed consent, on the other hand, is the map that leads us through the moral maze of using donor tissue. Just as you would ask permission before borrowing your neighbor's garden tools, scientists must obtain explicit consent from donors before using their cells in research or treatment.

Diving into resource allocation, we face the conundrum of limited treasure in a world of boundless need - a dilemma as old as shared crops in a small community. Decisions on which research projects are funded, and consequently, which therapies are pursued, often evoke debates as animated as town hall meetings.

Regulatory bodies and institutional review boards stand as the ethical guardians in this journey. Much like a council that upholds the rules of fair play in a community, these organizations ensure research maintains the highest ethical standards, respects individual rights, and aligns with societal values.

At times, the ethical roadmap of stem cell research seems to be in draft form, with discussions ongoing and consensus under construction. These dialogues shape the policy scaffolding of stem cell science just as public forums shape a city's landscape, with the power to lift its skyline or confine it to familiar contours.

In grasping the ethical dimensions of stem cell research, one begins to see the weighty responsibility of carrying the lantern of scientific inquiry along paths tread with respect, care, and an eye on the wider implications for humanity.

Technological advancements in stem cell research have catapulted the field into a new era of possibilities, particularly with the introduction of the CRISPR-Cas9 system. This groundbreaking tool functions like a tailor-made molecular scissor, allowing scientists to make precise alterations in the DNA sequence of a stem cell. By cutting and editing specific genes, researchers can investigate disease mechanisms more thoroughly or correct genetic defects directly at the source, offering potential cures where only symptomatic treatment was available before.

Not only has CRISPR-Cas9 accelerated the pace of stem cell research by simplifying the process of crafting genetically modified cells, but it also holds the key to unlocking therapies for conditions like sickle cell anemia and cystic fibrosis. Much like updating software to fix bugs, this technology enables the rewrite of genetic code to eliminate disease-causing errors.

However, as with any technology, limitations and ethical considerations exist. While CRISPR's capacity to edit genes is significant, ensuring these edits do not create unintended changes elsewhere in the genome is a challenge that researchers continue to navigate. The precision of CRISPR, the understanding of genetic interactions, and the effects of modifications on the human body are areas where ongoing research and debate indicate both caution and excitement for the future.

The CRISPR-Cas9 gene-editing process, when utilized with stem cells, is a sequence of intricate steps that offer a revolutionary approach to genetic alteration. Here's a detailed walk-through:

- Design of the Guide RNA (gRNA): Scientists first create a gRNA which serves as a navigation system, directing the CRISPR complex to the precise location of the gene that needs editing, much like a GPS provides a car with turn-by-turn directions.

- Homing to the Target DNA Sequence: The gRNA binds to the Cas9 enzyme, and this complex then travels along the DNA strand, searching for a matching sequence—a process akin to a detective looking for clues at a crime scene.

- Precise Cutting by the Cas9 Enzyme: Once the correct location is found, Cas9 acts like a pair of molecular scissors, cutting the DNA at exactly the right spot. Imagine a skilled artisan making an incision on a delicate piece of work—a stroke that must be perfect.

- Cell's Repair of the Cut: After the cut, the cell naturally attempts to repair the break. At this juncture, scientists can intervene to introduce a desired genetic change, in essence 'patching' the DNA as one would repair a garment, inserting a new segment or correcting a flawed one.

- Verification of Successful Gene-Editing: Scientists then examine the edited cells to confirm the DNA has been accurately altered. This is similar to a quality control process in manufacturing, ensuring the product meets the desired specifications.

To ensure the accuracy of CRISPR edits and reduce off-target effects (unintended cuts in the DNA), researchers use advanced software to predict potential gRNA binding sites. They also employ sequencing technologies post-editing to ensure only the target genes have been modified and no extraneous edits have occurred. This careful screening is akin to a proofreader meticulously checking a text for errors.

Moreover, researchers leverage integrated technologies such as machine learning algorithms to analyze and predict CRISPR outcomes before the actual editing. It's a bit like using weather prediction models before planning an agricultural season.

Current clinical trials are exploring the use of CRISPR-Cas9 in conjunction with stem cells for treating conditions such as sickle cell disease. However, challenges such as delivering the CRISPR components to the right cells in the body and ensuring long-term safety persist. These obstacles resemble the hurdle race in product development—from a concept to a well-tested and market-ready commodity.

Through ongoing research and clinical trial results, the potential of CRISPR-Cas9 in stem cell technology continues to crystallize, holding the promise for groundbreaking treatments that once seemed like the far-fetched dreams of science fiction.

International collaboration in stem cell research is akin to the combined efforts seen in constructing the International Space Station. Just as the ISS is a product of countries pooling their resources and expertise to create something that no single nation could accomplish alone, so too does stem cell research benefit from the cross-pollination of ideas and discoveries shared across borders. This cooperation allows for a broader range of genetic diversity in stem cell lines, vital for the study of diseases and the development of global treatments. Furthermore, diverse ethical perspectives and regulatory environments can enrich the debate and lead to more robust ethical frameworks. It's much like the way international aviation agreements ensure that, no matter where a plane is flying, safety and communication standards are upheld. By working together, scientists can accelerate the pace of discovery and increase the potential for life-saving therapies, showcasing the power of collaboration in our quest to understand and harness the possibilities of stem cells.

Here is the intricate weave of international stem cell research cooperation, detailed as if we're piecing together a global tapestry, each thread an essential part of the larger picture:

- **Structure of an International Collaborative Project:**
    - Initial Planning: Countries come together, much like chefs discussing the menu for a big event, to agree on the research focus and goals.
    - Defining roles based on strengths, like assigning who preps the starter or the dessert.
    - Establishing communication channels, like setting up a group chat for the culinary team.

    - Resource Sharing: Each country contributes resources, like each guest bringing an ingredient to a potluck dinner.
    - Labs and equipment access may be shared like kitchen space during a cooking class.
    - Funding is pooled from various sources akin to chipping in for a community feast.

    - Data and Information Exchange: Crucial information is exchanged freely and securely, similar to sharing recipes with trusted fellow cooks.
    - Databases are set up for data storage and analysis, like shared digital recipe books.
    - Regular meetings and conferences for updates, as if the cooks meet to taste and discuss each dish's progress.

- **Decision-Making Mechanisms:**
    - Steering Committees: Formed to make major decisions, like a council deciding town matters.
    - They set the research direction and protocols, akin to creating town guidelines.

    - Intellectual Property (IP) Management: Handled with care, ensuring all parties are acknowledged and benefit, just as co-authors are credited in a book.
    - Agreements on IP rights are forged, much like co-authors agree on royalties.

    - Distribution of Treatments: Agreed upon fairly, ensuring all countries involved gain access, similar to all the chefs at the end of a banquet receiving a portion of the leftover banquet to take home.
    - Guidelines for the distribution of any treatments much like rules for sharing the bounty of a community garden.

- **Regulatory Framework Integration:**

- National Regulations: Respected within the collaboration, like observing local customs when visiting another's home.
  - International standards are discussed and considered in the context of local laws, balancing global innovation with local norms.

  - Ethics Committees: Consulted to navigate the diverse ethical landscapes, akin to a council of elders upholding the community's moral fabric.

- **Real-World Example:**
  - The International Stem Cell Banking Initiative (ISCBI): Countries join forces sharing stem cell lines, much like libraries forming a network to exchange rare books.
  - They establish best practices for stem cell line derivation, storage, and distribution, ensuring consistency and quality as a librarian would for the precious books in their care.
  - This effort directly contributes to a greater understanding of stem cell applications and expedites the development of therapies, showing the fruit of combined labors on an international scale.

In essence, these collaborative projects in stem cell research are like global symphonies being composed—with each country's scientists, legal experts, and ethicists contributing notes to a harmonious and healing melody for humanity.

Imagine the long-term future of stem cell technology as humanity's next great voyage, akin to planning a colony on Mars. Just as astronauts might use local resources to build shelters on the Red Planet, future medical treatments could harness stem cells to repair damaged tissues right within our bodies. Envision a future where, instead of long-term medications for chronic conditions, a single stem cell treatment could provide a permanent fix, much like establishing a self-sustaining habitat in space. Each advancement in stem cell technology brings us closer to a day when managing a condition could mean a targeted, one-time treatment—pioneering a new norm in healthcare, just as space travel might one day become as routine as a transcontinental flight. It's about breaking the bounds of traditional medicine and venturing into a new horizon of healing, where the human body's own cells are the architects of recovery.

Let's take a deeper look at the advancements in stem cell therapy, which is akin to the evolution of personal computing from room-sized machines to the sleek smartphones we carry today. The progress has been remarkable, from the ability to grow organs from a patient's own cells, much like a gardener cultivating a rare plant from a tiny seedling, to targeting diseases with cell types crafted as precisely as a locksmith creates a key for a lock.

- **Organ Regeneration:**
    - Imagine a patient's damaged liver being restored using stem cells. Here, cells are harvested perhaps from the patient's skin or blood, then coaxed in the lab to become liver cells. It's like guiding an untrained voice to sing a complex aria, requiring expertise and a deep understanding of cellular behavior.

- **Precision in Targeting Tissues:**
    - Advanced techniques enable stem cells to be directed to become specific cell types. This can be compared to a master chef carefully selecting ingredients to achieve the perfect flavor profile for a dish, ensuring the final product blends seamlessly with its desired environment.

- **Reducing Immune Rejection:**
    - To reduce the chance of the body rejecting these new cells, researchers are fine-tuning the match between the patient's immune system and the lab-grown cells, similar to a tailor custom-fitting a suit for a client.

- **From Harvesting to Application:**
    - Harvesting cells from a donor or patient is the first step, akin to an artist gathering their paints and brushes. Subsequently, cells are cultured, molded into the needed form, and finally transplanted back to the patient - much like painting a masterpiece from scratch.

- **Overcoming Challenges:**
    - Researchers face hurdles such as scaling up this process to make it widely available, much like how a boutique manufacturer struggles to meet the demand for a suddenly trending product. They also work on ethically sourcing stem cells, which can be as complex and crucial as a jeweler ensuring that their diamonds are conflict-free.

- **Clinical Trials to Clinical Settings:**
    - Clinical trials are the test runs of stem cell therapy, much like an off-

Broadway show before it hits the main stage. These trials help fine-tune the therapies, determining safety and efficacy before they become standard treatment options. They pave the way for stem cell therapies to move from the benchtops of research labs to the bedside of patients.

As we peer into this future, stem cell therapies could revolutionize medicine in a way that makes once unfathomable treatments as common as downloading an app on our phone. The potential is immense, and with each breakthrough, the landscape of what's possible in medicine expands, promising a healthier, more resilient humanity.

Stem cell research stands at the forefront of medical innovation, holding the key to unlocking treatments for diseases previously deemed incurable. With their unique ability to differentiate into a variety of cell types, stem cells could potentially regenerate damaged organs, personalize medical interventions, and extend the frontiers of human health. The impact of this field extends beyond individual therapies to the transformation of healthcare systems globally. As research progresses, the promise of stem cells not only persists but expands—poised to redefine the boundaries of healing and enhance the quality of life for millions. This era of biomedical research underscores a pivotal shift towards targeted and regenerative treatments, cementing stem cells as a cornerstone of modern medical advancements.

CHAPTER 9 THE LAYMANS LABORATORY

Congrats, you've made it to Chapter 9: The Layman's Laboratory. Here, you'll take your inaugural steps into the realm of stem cell research. This chapter is tailored to simplify the intricate science of stem cells, unfolding it into tangible, hands-on activities that you can explore. Stem cells, the body's cellular building blocks with the remarkable ability to develop into diverse cell types, are integral to medical advancements and the healing of damaged tissues. Through this chapter, you'll learn how these powerful cells can be harnessed and observed right from the comfort of your home. This journey not only demystifies a crucial aspect of biology but also empowers you with practical skills and knowledge that contributes to a broader understanding of regenerative medicine.

Picture stem cells as the ultimate performers in the theater of the body, able to take on any role from brain cells to blood cells, much like an adept actor slipping into any character on stage. This is where Dr. Shinya Yamanaka's revolutionary work comes into play. Just as a skilled director can draw out the potential in an actor, Dr. Yamanaka discovered a way to prompt mature cells to revert to a versatile, embryonic state, ready to learn new parts—these are called induced pluripotent stem cells (iPSCs).

For the amateur scientist at home, this technique is like getting a masterclass from a renowned artist, enabling you to experiment with the very fundamentals of life. With instructions accessible as a simple script, anyone can explore the transformation of common cells into iPSCs. It means you, from your home laboratory, can partake in the kind of research that once required a proscenium of high-end lab equipment and advanced degrees. Dr. Yamanaka's work brings the science of possibility to your doorstep, unlocking the transformative performance of stem cells, offering a real-world glimpse into the regeneration and repair that these cellular wonders can conduct.

Dr. Shinya Yamanaka's landmark discovery in cell reprogramming is akin to teaching an experienced actor new roles in different genres of theatre. Here's a step-by-step guide to the process:

First, a mature cell, which might already have a specialized role, say a skin cell, is taken from the body. Think of this cell like an actor typecast in a specific part—limited to playing just that one role.

Next, the cell is introduced to a unique set of transcription factors, which are like scripts for entirely different roles it has never considered before. These transcription factors, named Oct3/4, Sox2, Klf4, and c-Myc, are pivotal in determining a cell's identity—essentially they re-script the cell's original role.

By introducing these factors, the cells are prompted to forget their former performances as skin cells. They now stand on a stage wiped clean, ready to take on a new script. Slowly, the cells revert to a pluripotent state, a blank slate, akin to an actor with an open future, no longer typecast and free to take on any role.

This pluripotent state is pivotal as it means these reprogrammed cells, now called induced pluripotent stem cells (iPSCs), have the potential to transform into any cell type in the body—from nerve cells to heart cells. This versatility is much like an actor skilled in comedy, drama, and action alike.

For the amateur scientist at home, recreating this process might sound like an ambitious production, but it's becoming increasingly feasible. While a home laboratory won't replicate the sophisticated environment of a research institution, it can still undertake related, simpler experiments that demonstrate the principles behind cell differentiation and pluripotency.

For example, you might not be able to recreate iPSCs on your kitchen table, but you can observe how plant cells differentiate in response to stimuli, without the need for specialist equipment. Placing a plant-cutting in water, you can watch roots develop from non-root cells, a clear demonstration of cellular potential driven by factors in their environment, much like the cellular reprogramming in Dr. Yamanaka's discovery.

Understanding each step of this process illuminates the profound implications it has for future healthcare and personalized medicine. It not only represents a thrilling scientific discovery but provides a framework for the realm of possibilities that stem cell research introduces—ranging from the study of diseases to the regeneration of damaged tissues. For the aspiring home scientist, it represents the first act in the exciting drama of cellular biology.

Setting up a home workspace for safe and efficient stem cell experiments is like preparing a cooking area before starting a complex recipe. First, you want to clear a dedicated space, free from clutter and any food items, to ensure a clean environment — think of it as your scientific countertop. It's crucial to have good lighting and access to a power source for any equipment you might need, similar to having a well-lit chopping board and an outlet for kitchen appliances.

Next, gather all the necessary materials and tools you will need within arm's reach, just as a chef assembles all ingredients before cooking. This might include items like gloves, safety goggles, a basic microscope, petri dishes, and so on. Organize these in an orderly fashion, perhaps in labeled containers or drawers, to avoid any potential cross-contamination — akin to separating raw meat from vegetables in the kitchen.

Safety is paramount. Keep a first aid kit nearby and familiarize yourself with the safety procedures for all your equipment. If you're working with any chemicals, even household items like vinegar or baking soda, treat them with respect and store them properly, as you would with sharp knives or a hot stove.

Remember, slow and steady wins the race. Never rush your experiments and always keep detailed notes, much like jotting down tweaks to recipes as you go. This meticulous approach not just builds good habits but also ensures that you can repeat your successes and learn from any mistakes, enhancing your journey into the world of amateur stem cell research.

Let's take a deeper look at the nuts and bolts of setting up a home lab for stem cell experimentation, rolling up our sleeves like a DIY enthusiast diving into a new project.

- **Procedural Checklist:**
  - Like prepping for a long road trip, kick off your lab setup with a go-to checklist:

- Confirm that you have personal protective equipment (PPE), like gloves and safety glasses—your travel gear for safety.
- Ensure a first-aid kit is close at hand, just as you'd keep a spare tire.
- Verify that your workspace is well-ventilated and a fire extinguisher is available, similar to setting up camp where you can breathe fresh air and douse a campfire if needed.

- **Risk Mitigation:**
- Like a meticulous gardener handling pesticides, manage biohazardous materials with care:
- Store all chemicals in clearly marked containers, much like keeping garden toxins out of reach of pets and children.
- Dispose of any biological waste properly, akin to how you would discard hazardous waste, not just tossing it into the compost pile.

- **Equipment Sterilization:**
- Think of your lab equipment like a set of kitchen knives that need to be cleaned before preparing a meal:
- Use a bleach solution or alcohol wipes to clean surfaces—your anti-bacterial soap for the lab.
- Sterilize instruments that will touch your samples, much like boiling baby bottles to keep them germ-free.

- **Data Collection and Documentation:**
- Approach this with the mindset of an avid birdwatcher, noting every detail in a journal:
- Record dates, times, and specific actions taken—scribble these down as if noting the comings and goings of rare birds.
- Photograph or sketch observations, just as you might capture the colorful plumage of a new bird species.

- **Troubleshooting Tips:**
- When unexpected results occur, channel your inner detective, much like Sherlock Holmes on a curious case:
- If contamination occurs, backtrack through your steps—trace your tracks to find where you may have gone off the path.
- When an experiment doesn't yield results, review your process and materials—re-examine your mystery novel for missed clues.

By following these guidelines, your kitchen table transforms into a stage for scientific discovery, allowing you to embrace the role of amateur scientist with gusto. And as you embark on this endeavor, every petri dish becomes a story, every microscope slide a world to explore, underlining the beautiful intricacy and boundless potential that stem cell science extends to us all.

Embarking on a basic stem cell experiment is much like following a recipe to craft a gourmet meal, where each step is essential and precision is key. Begin with your 'mise en place', gathering all the necessary 'ingredients' like stem cell cultures, growth mediums, and sterilized tools, ensuring everything is prepped and ready, just as you'd set out your herbs and spices before you start cooking.

The success of your dish depends on the precise measurement and timing of each ingredient, mirroring the methodical addition of nutrients and compounds to your stem cells. Monitor the temperature and conditions as you would watch over a delicate sauce simmering on the stove. As in cooking, adjusting the 'flavors' of the cell environment can lead to a strikingly different outcome, guiding the cells to develop into the desired 'dish' or tissue type.

Throughout the process, keep detailed notes of every step in your 'recipe book', documenting your method as meticulously as a chef perfecting a new dish. Remember, just as the best chefs taste and adjust their creations as they go, regularly observe your cells, adapting your technique to ensure they 'cook' to perfection.

In the end, the dish — or in this case, the newly specialized cells — should be a testimony to your careful preparation and attention to detail, showcasing the incredible potential of stem cells when nurtured in just the right way. This journey through the kitchen of science will not only satisfy your intellectual appetite but also serve as a springboard for further culinary — and scientific — explorations.

Here is the step-by-step breakdown of setting up a stem cell experiment, presented as you would go about crafting a complex and delightful dish in your kitchen:

Preparing the Workspace:
- Clear the counter, ensuring that the area is uncluttered and clean – just

as you would need a tidy surface for rolling dough.
- Sterilize the workspace using disinfectants – think of it like using a food-safe cleaner before you start cooking.

Sanitizing Tools:
- Gather lab tools such as pipettes, tweezers, and petri dishes – akin to assembling your measuring cups, knives, and mixing bowls.
- Sterilize these tools using an autoclave or appropriate sanitizing solution, much as you would ensure your cooking utensils are germ-free before use.

Obtaining and Handling Stem Cells:
- Acquire stem cells or prepare them from an existing culture, similar to sourcing the freshest ingredients from the market or your pantry.
- Handle cells with care under sterile conditions to prevent contamination, much like keeping raw and cooked foods separate to avoid cross-contamination.

Monitoring Environmental Conditions:
- Maintain a consistent temperature suitable for stem cell growth, as you would keep an oven at a steady heat for baking.
- Monitor pH levels of the culture medium, analogous to checking the seasoning levels in a dish you're preparing.
- Keep track of the duration of the experiment, setting timers as you would to remind you when to check on your soufflé.

Adjusting Variables:
- If cell growth is not proceeding as expected, adjust the growth medium or environmental conditions – similar to tweaking a recipe when the flavor isn't just right.
- Be patient and methodical in making adjustments, avoiding drastic changes that could 'spoil the dish.'

Troubleshooting Common Issues:
- If contamination occurs, evaluate your sterile technique and make adjustments – think of salvaging a sauce that's started to curdle.
- For culture inconsistencies, examine the uniformity of your cell seeding technique, akin to ensuring even distribution of batter in a cake pan.
- Keep a log of all actions and results, as a baker would note variations in

a recipe that led to the perfect loaf of bread.

By following these detailed steps, conducting a stem cell experiment can be a manageable and rewarding 'culinary' journey in the lab. With each careful measure and observant eye, you become both the chef and the scientist, capable of crafting something that is as beautiful as it is scientifically significant.

The process of documenting and analyzing data from stem cell experiments is critical—think of it as the investigative work that follows the excitement of conducting the experiment itself. Start by observing and noting the appearance, behavior, and changes in cells at regular intervals, just as a bird watcher keenly records sightings in a logbook. Precision here is key, and the most subtle changes can be significant, much as a shifting breeze can signal changing weather.

Equally, recording numerical data such as cell counts, growth rates, and temperature readings with rigor is essential; it's the scientific equivalent of a financial analyst keeping meticulous records of market movements. Utilize spreadsheets or lab software to track this information systematically—tools that serve as the ledger in your scientific accounting.

When it's time to make sense of the data, approach it with a detective's acumen. Look for patterns as you would clues, asking questions and formulating hypotheses. If a particular set of cells is thriving, consider what conditions differed from less successful trials. Is it akin to discovering which soil yields the best garden growth, prompting further exploration and understanding?

Occasionally, the data may not align with your expectations. Rather than discard these as errors, examine them as a mechanic might a misfiring engine—what can they tell you about the process that isn't immediately obvious? These anomalies could lead to new insights, pushing the boundaries of what is currently known.

Concluding the analysis, consolidate your findings into a cohesive report or a set of conclusions. Imagine crafting a narrative that explains the journey your cells have undergone. This documentation is a vital component of the scientific method, just as a well-prepared financial report can illuminate the performance of a business. It not only serves as a record of your work but can be a guide for others to follow and build upon, laying one more brick in the collective edifice of scientific knowledge.

Let's take a deeper look at the meticulous process of data collection in stem cell experiments, akin to a detective sifting through clues at a crime scene.

Quantitative Measurements:
- Start with precise numbers just as a carpenter measures wood before cutting. This includes cell count, viability percentages, and growth rates. You'll need tools like a hemocytometer for counting and software for tracking growth curves over time.

Qualitative Observations:
- These are your notes on the cells' look and behavior — much like critiquing the style and performance of an actor. Under the microscope, observe differences in cell shape, size, and movement. It's visual storytelling that offers clues about cell health and development.

Control Variables:
- Think of these as your recipe constants, like always using the same oven temperature when baking bread to ensure consistent results. Control variables in your experiments might include the type of culture medium used, incubation times, and room conditions, such as light and humidity.

Analyzing Data:
- Understanding cell morphology and growth patterns is like a gardener recognizing plant health and growth stages. Analyze cell images for indicators of differentiation or abnormal growth, considering these insights the 'fruits' of your scientific garden.

Documenting Variations:
- Record any changes to protocols thoroughly in your lab notebook — your culinary journal for scientific experiments. Document the time, date, and nature of variations, just as you would note alterations to a recipe and the effects on the final dish.

Control Groups:
- These are your baseline comparisons. Imagine trying a new fertilizing method on one plant while leaving another untreated. The untreated plant serves as your control, a standard against which all other results are measured.

Handling Discrepancies:
- When reality doesn't match your hypothesis, don't sweep it aside like an unsightly stain. Instead, analyze it as you would solve a puzzle, identifying all potential influences, from cross-contamination to an equipment malfunction.

Systematic Testing of Impact:
- Systematically test each variable that could affect your results, like troubleshooting why your cake didn't rise and methodically testing each ingredient and step. In the lab, this might look like adjusting one condition at a time and observing the effect.

Engaging with data in this way transforms you from a passive observer to an active participant in the narrative of discovery. Like piecing together a complex plot in a novel, your attention to detail sets the stage for breakthroughs and deepens your connection to the world of science, making every data point a character in the unfolding story of stem cell research.

Henrietta Lacks' story is a seminal case in understanding the importance of ethics in research. Without her consent, Lacks' cells were used to create the first immortalized human cell line, contributing significantly to scientific advances, including the development of the polio vaccine. This highlights the critical role of informed consent and respect for a person's bodily autonomy in research. Handling biological materials, researchers must prioritize ethical considerations, ensuring transparency and obtaining explicit consent. Henrietta Lacks' legacy teaches us the value of ethical standards in respecting individual rights while advancing medical science. It serves as a reminder to handle biological specimens with the utmost care, acknowledging their origin and the individuals from whom they come.

Just as a cherished recipe passed down through generations gains new flavors and tweaks with each cook, documenting and sharing your scientific experiments can enrich the collective knowledge pool. Imagine each step of your experiment as an ingredient in a recipe; by writing down the precise measurements, timing, and techniques, you create a guide that others can follow and build upon. Sharing this 'recipe' with fellow science enthusiasts is like contributing to a community cookbook; your unique approach may just hold the key to a breakthrough another researcher has been seeking. As everyone adds their findings, the community thrives, learning not only from successes but also from the proverbial sunken cakes and over-seasoned stews. This exchange turns isolated experiments into a shared scientific feast, where every researcher, regardless of experience level, can both teach and learn, dine and contribute at the table of discovery.

Becoming a citizen scientist in the field of stem cell research is an empowering journey that transforms complex biological concepts into accessible knowledge. As individuals engage with the hands-on study of stem cells, they gain invaluable insights into cellular mechanisms that shape health and disease. This endeavor not only satisfies personal curiosity but also contributes to a larger scientific dialogue. By demystifying the nuances of stem cells, citizen scientists can influence the trajectory of research, potentially leading to new discoveries. The aggregate impact of their meticulous work and shared experiences amplifies the progress of stem cell science, making it a collective effort where each contribution is a step forward in understanding and innovation.

CONCLUSION

As we draw the curtain on our exploration of 'Stem Cells Made Easy,' we reflect on a journey that has spanned the microscopic to the macroscopic, traversing the complex landscape of stem cell science with guidance from vivid analogies and real-world applications. We've unraveled the mysteries of these potent cells, understanding not only their biological significance but also their transformative potential in medicine.

Key themes have emerged from the chapters within: the remarkable plasticity of stem cells, the ethical considerations that guide their use, and the burgeoning technological advances that are revolutionizing healthcare. We've seen how stem cells can behave like the master keys of the body, unlocking healing and regeneration where once there was damage and disease.

The lessons learned stretch beyond the scientific facts and into the realm of possibility. We stand at the threshold of a new era in medical science, where personalized medicine could become a reality, and ailments once deemed terminal could be conquered. We've learned that stem cells are more than just a biological wonder; they represent hope for countless individuals awaiting breakthroughs in treatment and cures.

The impact of this book may resonate differently with each reader, but the collective takeaway is a greater appreciation for both the intricacies and the ethical considerations of stem cell research. We depart from this textual odyssey equipped with knowledge, stirred by the potential of future innovations, and mindful of the responsibility that comes with such powerful science.

Let us ponder on the path ahead for stem cells, not just as a subject of scientific curiosity but as a beacon of progress. The trials, triumphs, and challenges that lie in the interplay between research and application ensure that the story of stem cells will continue to evolve, and with it, so will our understanding. May this book serve as a foundation for your continuous journey into the remarkable realm of stem cells—a journey marked by constant learning, awe, and wonder at the capabilities tucked within the building blocks of life.

ABOUT THE AUTHOR

Jon Adams brings a wealth of experience from over twenty years in the information technology industry, having worked with some of the world's leading tech giants. With a deep-seated passion for science, technology, and languages, Jon excels at demystifying complex subjects, making them accessible and engaging to a broad audience. His writings focus on breaking down intricate topics into everyday terms, helping readers not just learn but also apply this knowledge in their daily lives.

Currently, Jon is a proud member of Green Mountain Computing, which publishes his insightful books. Through his work, he aims to foster a deeper understanding and appreciation of technology and science, enriching readers' lives.

Jon@GreenMountainComputing.com

www.ingramcontent.com/pod-product-compliance
Lightning Source LLC
Chambersburg PA
CBHW070303230526
45470CB00002B/693